Atlas of igneous rocks and their textures

Atlas of igneous rocks and their textures

W.S. MacKenzie, C.H. Donaldson and C. Guilford

Addison Wesley Longman Limited
Edinburgh Gate,
Harlow,
Essex CM20 2JE, England
and Associated Companies throughout the world

First published 1982
Second impression 1984
Third impression 1987
Fourth impression 1991
Fifth impression 1993
Sixth impression 1995
Seventh impression 1997

British Library Cataloguing in Publication Data
MacKenzie, W. S.
Atlas of igneous rocks and their textures.
1. Rocks, Igneous—Pictorial works
I. Title II. Donaldson, C. H.
III. Guilford, C.
552.1′0222 QE461
ISBN 0-582-30082-7

Library of Congress Cataloging in Publication Data
A Catalog entry for this title is available from the Library of Congress.

Set in 9/10 pt. Monophoto Times New Roman
Produced by Longman Asia Ltd, Hong Kong
SWTC/07

Contents

Part 2 Varieties of igneous rocks
(Numbers refer to photographs – not to pages)

Intermediate rocks

Acid rocks

Alkaline and miscellaneous rocks

Preface

The commonest means of studying an igneous rock is to examine it in thin section, either with a petrographic microscope or a hand lens, which permits identification of the minerals present and investigation of their textural relations. From such study the skilled petrographer can interpret details of the history of the magma which crystallized to form the rock.

To become skilled requires many hours of study and training. Much of the training is acquired by patient attention by the teacher to the student. The student needs his observations verified and this can result in the teacher being summoned every minute or so; with a class of ten or more, the student is for long periods unattended, becomes frustrated and loses interest. The remedy is for the student to be able to verify his own observations by comparison with a photograph of a rock of the same type or showing the same feature(s).

The main aim of this book is to provide such a laboratory handbook to assist the student of geology (undergraduate and amateur) beginning to study igneous rocks in thin section. It is hoped that it may also be useful as a reference work for more advanced students and others interested in the natural history of rocks.

The work is divided into two parts – Part 1 is devoted to descriptions and photographs of textures found in igneous rocks and Part 2 consists of photographs of common (and a few not so common) igneous rocks.

We have selected those rocks and textural types which we believe may be encountered in an undergraduate course in geology but have made no attempt to produce a comprehensive coverage of all igneous rocks which have been given individual names because many of these names reflect only minor mineralogical or textural differences.

We have tried, as far as possible, to avoid any interpretation of the origin of textures and rocks, although the simple matter of arranging the rocks in some order of presentation is based to some extent on presumed genetic relationships between them.

In a previously published *Atlas of Rock-Forming Minerals* we have illustrated the appearance of the common rock-forming minerals so that here we have not considered it necessary to describe the optical properties in detail. To be able to give a name to the majority of igneous rocks it is only necessary to be familiar with the properties of between twelve and fifteen minerals and we have assumed that the user of this book is already able to recognize these minerals.

Thin sections can be observed under the simplest of microscopes fitted with two pieces of polaroid and a new field of interest is open to the amateur for only a modest financial outlay. Because some amateur geologists may be interested in preparing their own thin sections we have included a brief description of how this may be done.

Many of the photographs show a combination of shapes and colours which have a special beauty of their own, reflecting the fact that while thin section study is of practical importance it can also be of aesthetic satisfaction. A few of the most attractive pictures unashamedly represent the authors' self-indulgence.

The usual criticism of photomicrographs is that it is very often difficult to determine which feature they are intended to illustrate. For this reason many textbooks are illustrated by drawings in which the required feature may be exaggerated. We have tried to test the usefulness of our photographs by asking our colleagues to identify the mineral assemblage or texture which we have illustrated. We are grateful to them for their help in this respect. Most of the photographs were made from thin sections of rocks in the teaching collections of the Geology Departments of Manchester University and St Andrews University. Others were provided by friends and colleagues who made available to us thin sections from their own research collections, and we are most grateful to them for their help in this matter. We are particularly indebted to Dr John Wadsworth and Mr Ian MacKenzie who read and criticized all the descriptions of the textures and rocks. However, any failings in these descriptions are our responsibility alone.

Finally, we caution those using the book not to regard the photographs as representing all the known textures and varieties of igneous rock, or indeed all their guises. These photographs are only an *aid* to recognition of textures and rock types and can never substitute for looking at thin sections under the microscope.

Acknowledgements

We are much indebted to our colleagues and friends who have generously given us thin sections of rocks from which to take photographs: they include the following gentlemen; S. O. Agrell, B. Atkinson, N. Binstead, K. Brooks, F. M. Broadhurst, I.S.E. Carmichael, J. B. Dawson, J. Esson, M. E. Fleet, F. G. F. Gibb, A. Hall, D. L. Hamilton, C. M. B. Henderson, A. M. Hopgood, E. Iki, R. Johnston, I. R. MacKenzie, R. Nesbitt, E. Sapountzis, J. Wadsworth, Rong-shu Zeng and J. Zussman. We have also benefited greatly from having beem able to use the collections of the late Prof. H. I. Drever who was the teacher and friend of both WSM and CHD at St. Andrews University, although 25 years intervened between our time as his students: he and his colleague Mr. R. Johnston were jointly responsible for arousing our interest in igneous rocks early in our careers.

The staff of the publishers have been very patient and helpful and we especially wish to thank them for their consideration and for that quality essential to all publishers – a sense of humour.

Miss Patricia Crook's help both in typing the manuscript and in preparing the index is gratefully acknowledged. We are grateful to Dr Robert Hutchison of the British Museum (Natural History) for permission to photograph thin sections of the Prairie Dog meteorite and the Stannern meteorite, both of which are in the British Museum collections.

Part 1

The textures of igneous rocks

Introduction

To English-speaking petrologists *textures* are the geometrical relationships among the component crystals of a rock and any amorphous materials (glass or gas in cavities) that may be present. They comprise the following properties:

1. Crystallinity (degree of crystallization) – i.e. the relative proportions of glass and crystals.
2. Granularity (grain size) – i.e. the absolute and the relative sizes of crystals.
3. Crystal shapes.
4. Mutual relations or arrangement of crystals and any amorphous materials present.

In this part of the book textures in each of these categories are described and illustrated, some in plane-polarized light (PPL), some in cross-polarized light (XPL) and some in both. Some textures exhibit more than one of the above properties and we have indicated where this is so.

Petrography, of which textural relations are a part, is the descriptive and factual side of *petrology*, whereas *petrogenesis* is the interpretive side. Thus genetic terms, such as *cumulate*, *cumulus crystal*, *cumulate texture*, *synneusis texture*, *exsolution texture* and *fluxion texture* should be avoided, as they combine factual description with interpretation; they rob any person reading a petrographic description of unbiased observations and can cast doubt on the objectivity of the petrographer who wrote the description. For this reason, genetic textural terms are not included in this book, there being suitable non-genetic terms available for all of them.

Remarkably few igneous textures have been reproduced in the laboratory and the origins of even fewer could be claimed to be adequately understood. For these reasons, we have made no comment on the origin of most of the textures; readers should consult the texts by Iddings (1909), Holmes (1921), Niggli (1954), Hatch, Wells and Wells (1972) or Cox, Bell and Pankhurst (1979), for discussion of the origin of textures and their implications. However, it should be noted that many textures are open to more than one interpretation and the newcomer to the subject is advised to consider the possible origins and implications for himself before reading one of these texts. He is then likely to interpret the crystallization of a rock more objectively and flexibly than if the 'standard interpretation' is adopted slavishly. This comment is particularly relevant to the interpretation of 'order of crystallization' of minerals in a rock. We have found that both students and teacher can benefit from a two-hour discussion of the subject; the student who is unencumbered by preconceptions can be remarkably inventive and provide his teacher with copious new ideas for consideration.

In studying rocks in thin section we must not forget that only a two-dimensional view is present and hence the true three-dimensional texture has to be deduced from examination of the dispositions of many crystals in the section. In rocks with a strong preferred orientation of crystals, two or more sections of different attitude may be required to reveal the texture adequately.

Crystallinity

Igneous rocks range in crystallinity from entirely crystals to entirely glass. Adjectives used to describe these states are shown on the following scale:

100% crystals		*100% glass*
holocrystalline	*hypocrystalline*[1] or *hypohyaline*	*holohyaline*

The adjectives *glassy*, *vitreous* and *hyaline* all indicate that a rock is more or less completely glass.

[1] *Hypocrystalline rocks can be described more precisely by stating the relative proportions of crystals to glass.*

1 Holocrystalline anorthositic gabbro

Elongate crystals of plagioclase feldspar, some wrapped round olivine crystals, form a framework in this rock, the interstices of which are filled with smaller plagioclase, olivine and augite crystals. The purplish-blue area at the top right of this photograph is an augite crystal which includes a number of small plagioclase and olivine crystals.

Perpendicular Feldspar gabbro from Middle Border Group of the Skaergaard intrusion, East Greenland; magnification ×7, XPL.

2 Holocrystalline granite

Crystals of biotite, quartz, 'perthitic' potassium-rich feldspar (large crystal bottom right) and zoned sodium-rich feldspar makes up this granite. The speckled appearance in the cores of the plagioclase feldspars is caused by fine inclusions of mica.

Granite from Ross of Mull, Scotland; magnification ×14, XPL.

3 Hypocrystalline pitchstone with perlitic cracks

Crystals of plagioclase, biotite and magnetite in this rock are set in glass (black in XPL) which has spherical fractures known as *perlitic cracks*: these appear as circles in thin section.

Dacite from Chemnitz, East Germany; magnification × 20, PPL and XPL.

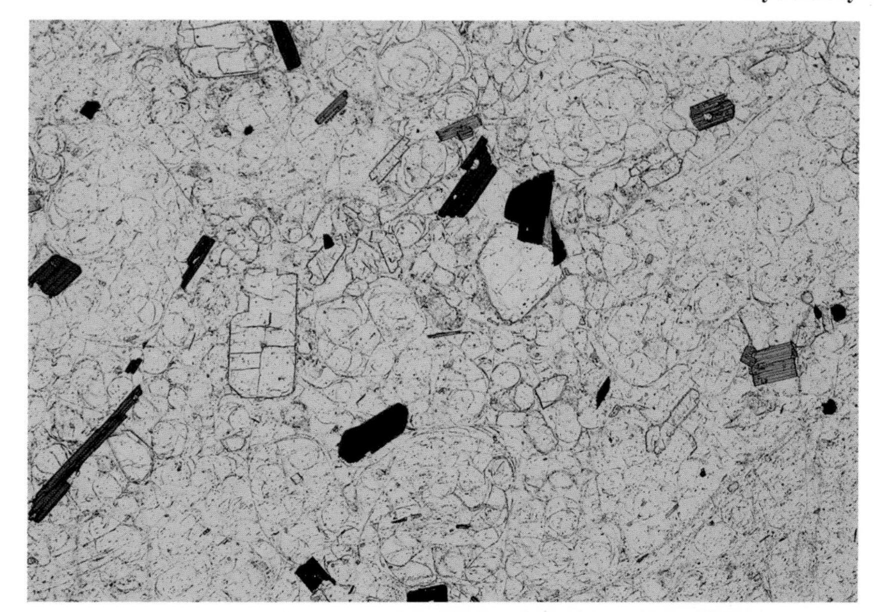

4 Hypocrystalline basalt

Small olivine phenocrysts (colourless in PPL) and columnar, skeletal titanaugite crystals (pinkish-beige colour in PPL) are enclosed by murky brown glass. No plagioclase has crystallized in this rock. The deeper pink colour around the margin of some of the titanaugites is a narrow mantle of Ti-rich amphibole.

Basalt from Quarsut, West Greenland; magnification × 35, PPL and XPL.

Hypocrystalline basalt (continued)

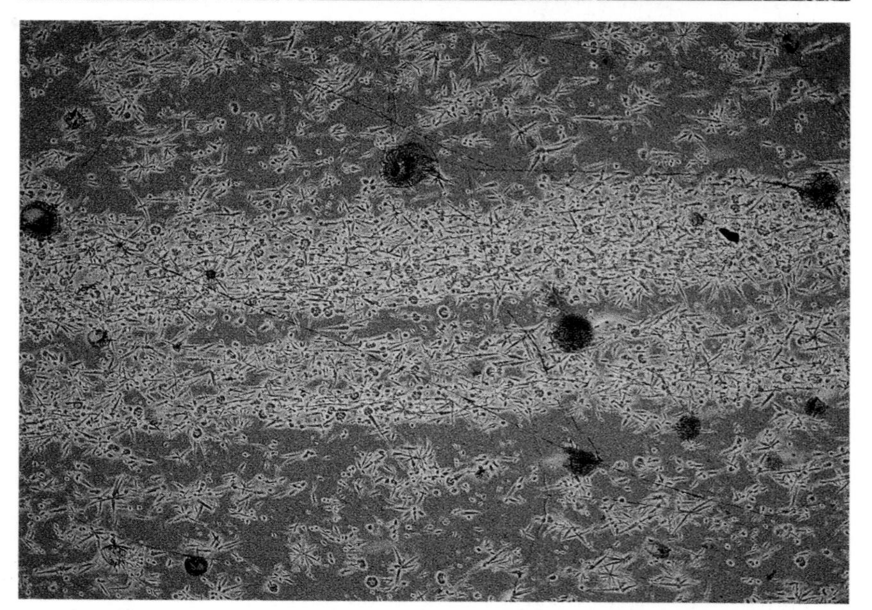

5 Glassy rock

The photograph shows abundant, very small crystals (probably quartz or feldspar) enclosed by glass. Note the banding caused by (a) differences in abundance of crystallites, (b) crystallites in the lighter bands having a slight preferred alignment and (c) differences in colour of the glass. The small brown, isolated round objects are known as 'spherulites' (see *Spherulitic texture*, p. 54). (See also **14**.)

Pitchstone from Arran, Scotland; magnification × *12, PPL.*

6 Glassy basalt threads – Pele's hair

These filaments of basalt glass form when particles in a molten lava spray are caught by the wind and drawn out. Pele is a mythical lady, believed by native Hawaiians to reside within the volcano Kilauea. (Contrast **7**.)

Specimen from Erta Alé volcano, Ethiopia; magnification × *8, PPL.*

7 Glassy particles of mare basalt in lunar soil

Pieces of glass, many of them spherical, are orange-brown or black in colour. Some of the darker ones are partially crystalline. These particles were formed by rapid cooling of droplets of basalt melt; it has been suggested that the droplets formed either in a fire-fountaining lava eruption, or by meteorite impact into a lava lake or into a molten or solid lava flow. (Contrast **6**.) The scarce, irregularly shaped fragments are pyroxene (pale brown) and feldspar (colourless).

Lunar basalt 74220 from Taurus Littrow Valley collected by Apollo 17 astronauts; magnification × 43, PPL.

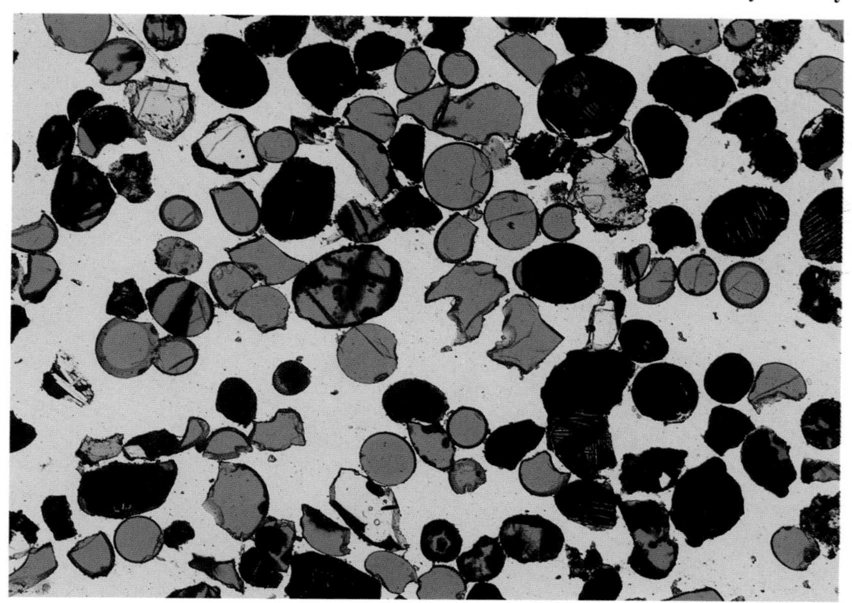

7a Liquid Immiscibility

Globules of one glass in another are found in some rocks and these are attributed to immiscibility of the two liquids. In this rock they can only be seen at very high power in thin films of glass between laths of plagioclase.

Specimen from basalt lava, Lava beds National Monument California, U.S.A.: magnification × 600, PPL.

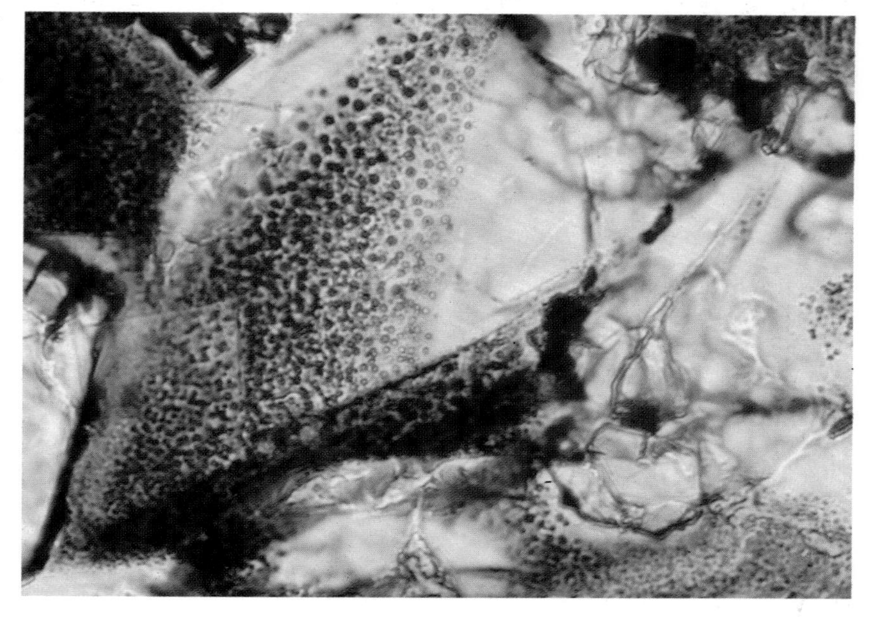

Glass, or devitrified glass, is often an important constituent of the pyroclastic rocks known as *ash-fall tuffs* and *ash-flow tuffs* (or *ignimbrites*). Such rocks typically have *fragmental textures*, i.e. they comprise mixtures of fragments of rocks, crystals and glass, predominantly less than a millimetre in size (**8–9**). In an ash-flow deposit the glass fragments may initially be plastic enough to be partly or wholly welded together as the weight of overlying material causes compaction of the constituent fragments; such a rock is known as a *welded tuff* (**8b**). If sufficient heat is available, glassy fragments devitrify.

8a Glassy unwelded rhyolite tuff

The glassy fragments in this rock, some of which are banded and slightly flattened, are not welded to one another. They and the crystals of quartz and feldspar are embedded in fine glassy particles (ash).

Recent ignimbrite, from Whakatane, North Island, New Zealand; magnification × 46, PPL

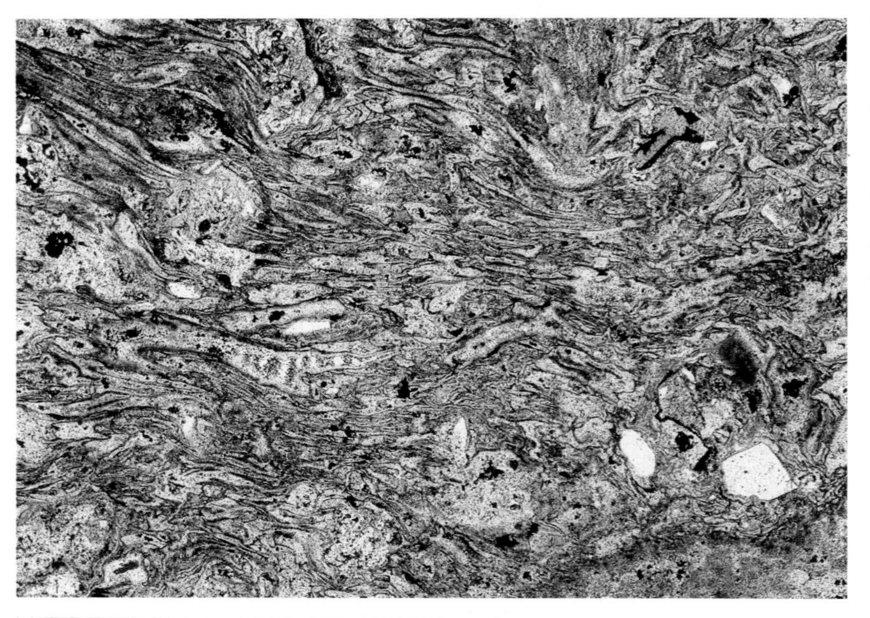

8b Glassy welded crystal tuff

The glassy matrix in this rock has an apparent discontinuous lamination caused by extreme compaction and welding of original pumice fragments. The regular alignment of the flattened fragments is known as *eutaxitic texture*.

Welded tuff from Tibchi granite ring-complex, Nigeria; magnification × 36, PPL.

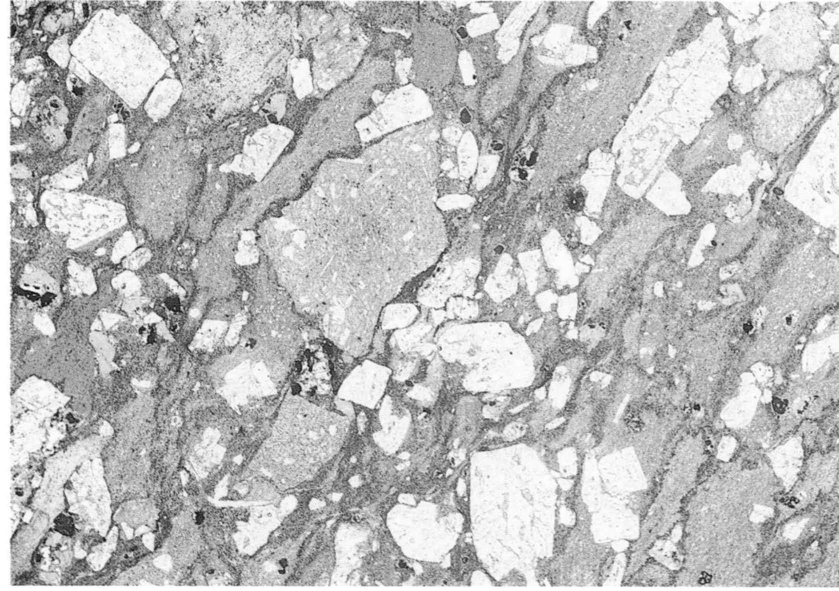

9 Tuff

This fragmental rock consists of crystals of quartz, alkali feldspar and plagioclase of various sizes and shapes, pieces of glassy rhyolite (e.g. centre) and pieces of fine-grained tuff, all enclosed in a fine-grained banded ash matrix which originally may have been glassy. (See also **13**.)

Tuff from Llanellwedd, Wales; magnification × 10, PPL and XPL.

Granularity

This property embraces three different concepts: (1) what the aided and unaided eye can or cannot see; (2) absolute crystal sizes (p. 12); and (3) relative crystal sizes (p. 14).

Terms referring to what the aided and unaided eye can or cannot see

Phanerocrystalline (*phaneritic texture* of American petrologists) – all crystals of the principal minerals can be distinguished by the naked eye (see **10**).[1]
Aphanitic – all crystals, other than any phenocrysts present (see p. 14), cannot be distinguished by the naked eye.[2] Two sub-types exist:

(a) *Microcrystalline* – crystals can be identified in thin section with a petrographic microscope (**11**). Crystals only just large enough to show polarization colours (less than 0.01mm) are called *microlites*.
(b) *Cryptocrystalline*[3] – crystals are too small to be identified even with the microscope (**12** and **13**). Globular, rod-like and hair-like crystals which are too small to show polarization colours are known as *crystallites*.

[1]Pegmatitic texture *is a variety of phanerocrystalline in which the crytals are strikingly large, bigger than 1–2 cm, and in rare instances up to many metres.*
[2]*The term* aphyric *is sometimes used for aphanitic rocks which lack phenocrysts* **(eg, 60, 63, 107)**.
[3]Felsitic texture *is sometimes applied to siliceous rocks with ill-defined, almost cryptocrystalline, grey-polarizing areas composed of more or less equigranular aggregates of quartz and alkali feldspar. The name* felsite *is often applied to such rocks, although this is more commonly a field term for fine-grained acid material of uncertain mode off occurrence.*

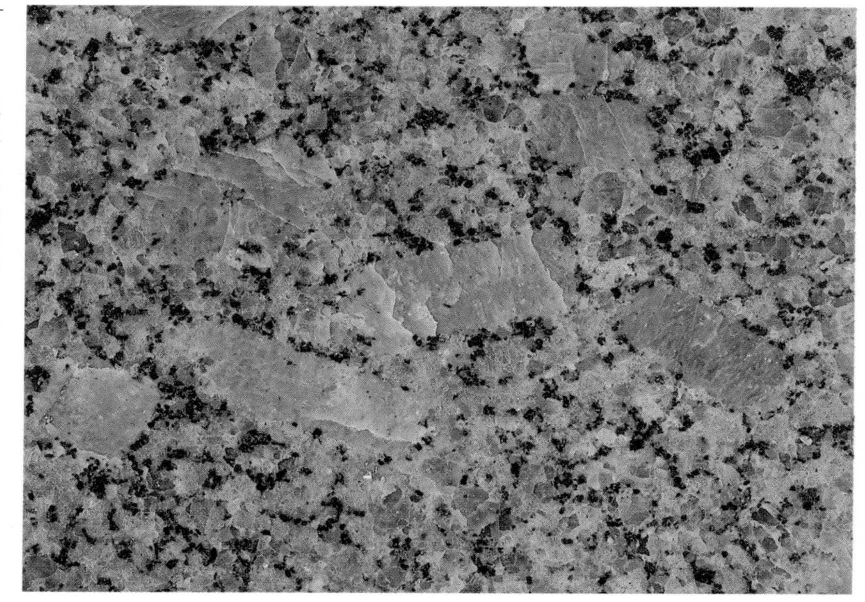

10 Phanerocrystalline granites

The crystals in the two granites, illustrated here in hand specimen, are clearly visible to the naked eye. Although the rocks contain the same minerals (alkali feldspar, plagioclase feldspar, quartz and biotite) the proportions of the minerals are not the same, and this influences the rock textures. Thus the Shap granite contains two distinct sizes of potassium feldspar crystals (pink), whereas the Eagle Red Granite has only one.

Granite from Shap, England (opposite) and 'Eagle Red' granite, South Africa (next page); both magnifications × 1.

A thin section view of the Shap granite is shown in ***144***

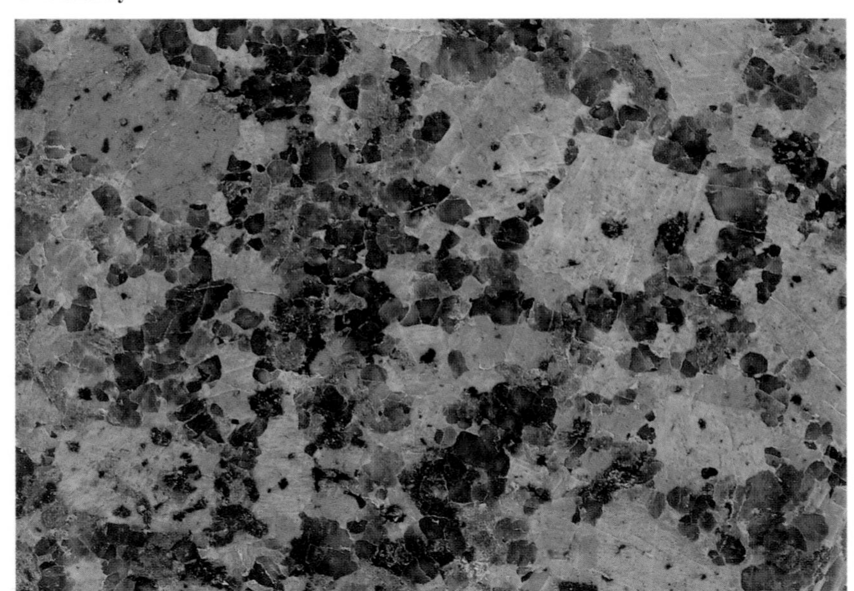

Phanerocrystalline granites (continued)

11 Microcrystalline olivine basalt

This rock consists mainly of plagioclase feldspar, augite and olivine but, without the aid of the microscope, individual crystals would not have been distinguishable. In parts of the photograph the randomly arranged rectangular plagioclases are enclosed by areas showing uniform yellowish interference colours, these are augite crystals.

Olivine basalt from North-west Skye, Scotland; magnification × 11, PPL.

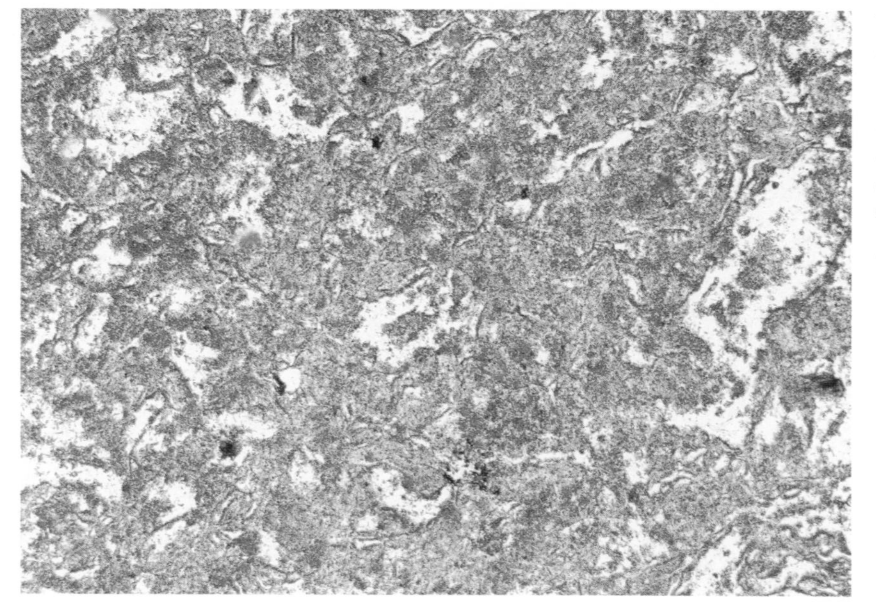

12 Cryptocrystalline rock

Comparison of these two photographs shows that the brown material in the PPL view is birefringent but that the individual crystals are of submicroscopic size. The clear areas in the PPL view are slightly more coarsely crystalline, as can be seen in the XPL view.

Rhyolite from Island of Pantelleria, Italy; magnification × 72, PPL and XPL.

Cryptocrystalline rock (continued)

13 Cryptocrystalline matrix in a tuff

Cryptocrystalline texture is common in tuffs (i.e. consolidated ash), as in the matrix of this rock. Here the matrix encloses fragments of shale and quartz crystals. (See also **8** and **9**.)

Tuff from unknown locality; magnification × 16, PPL and XPL.

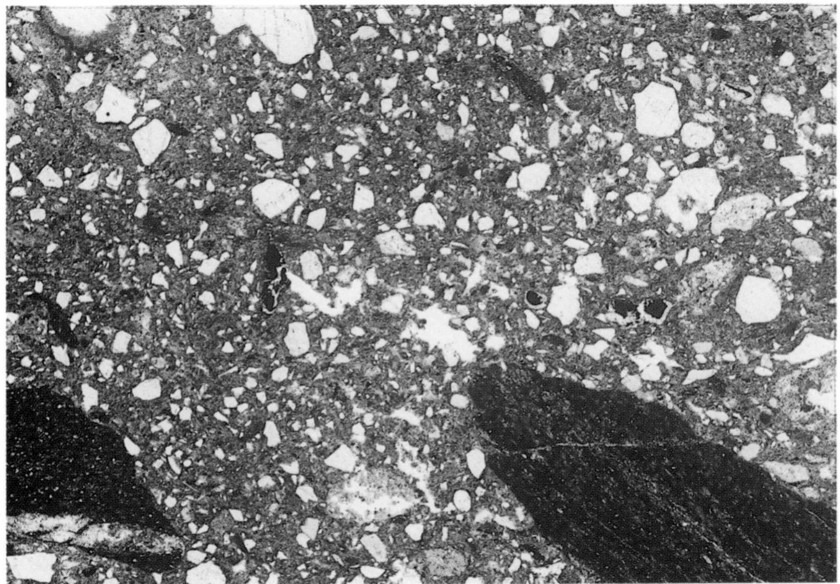

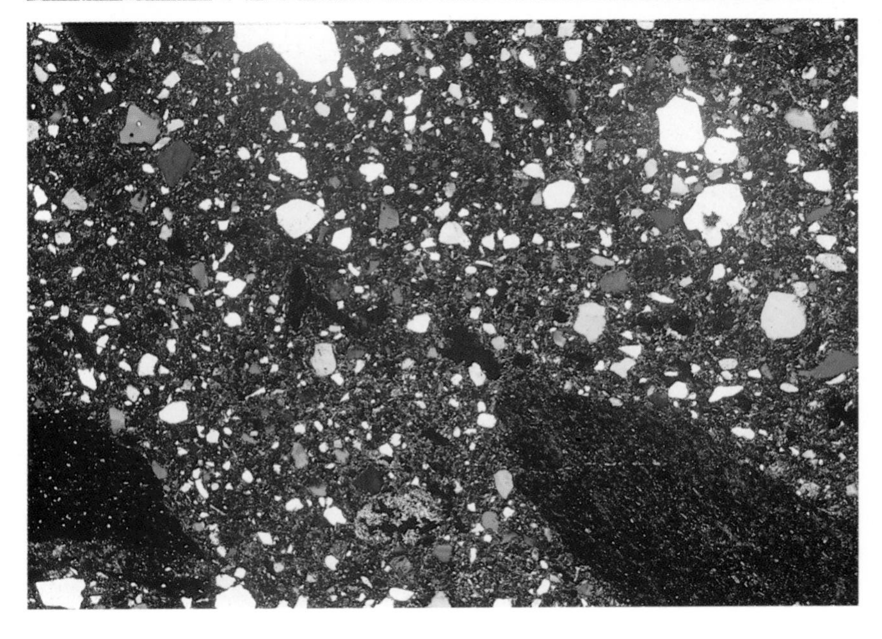

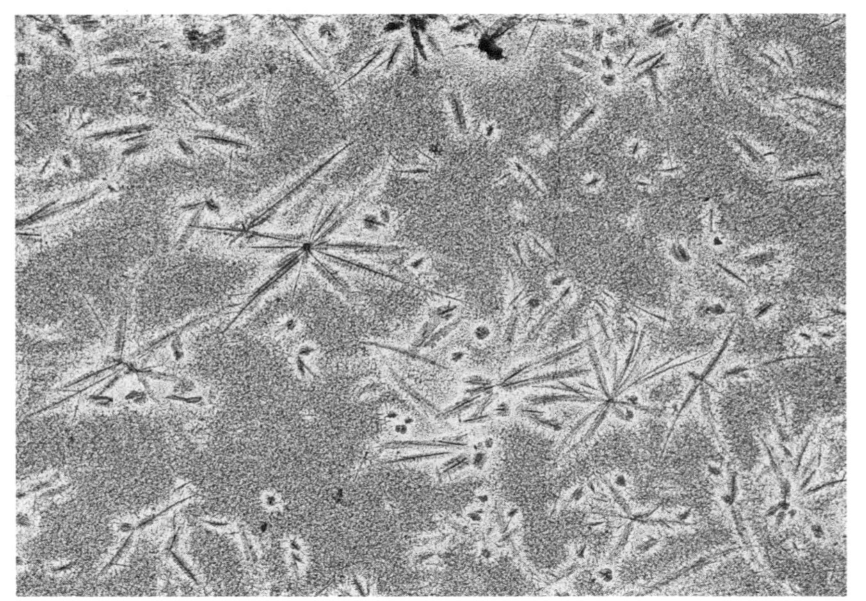

14 Pitchstone containing crystallites of two sizes

Radiate clusters of crystallites are set here in glass. The bulk of the glass contains even smaller crystallites, causing the grey colour, whereas adjacent to the larger crystallites the smaller ones are absent. This is a higher magnification view of the rock illustrated in **5**.

Pitchstone from Arran, Scotland; magnification × 52, PPL.

Terms indicating absolute ranges of grain size

Coarse-grained – crystal diameters > 5 mm
Medium-grained – crystal diameters 1–5 mm
Fine-grained – crystal diameters < 1 mm[1]

The next six photographs (**15**, **16** and **17**) were all taken at the same magnification (× 27) to indicate how grain size relates to the number of crystals seen in a given field of view (4.2 × 3.1 mm), and hence the extent of the texture visible at that magnification. While the overall texture is recognizable in the fine-grained rock, it is not so in the coarse one and a low-power objective lens would be necessary to examine it adequately. Petrographic microscopes rarely have a sufficiently low-power objective lens for examining the textures of coarse-grained rocks; a hand lens should be used for these, with two sheets of polaroid, if available.

[1] *Some petrologists include another range, < 0.05 mm, which they call* very fine-grained.

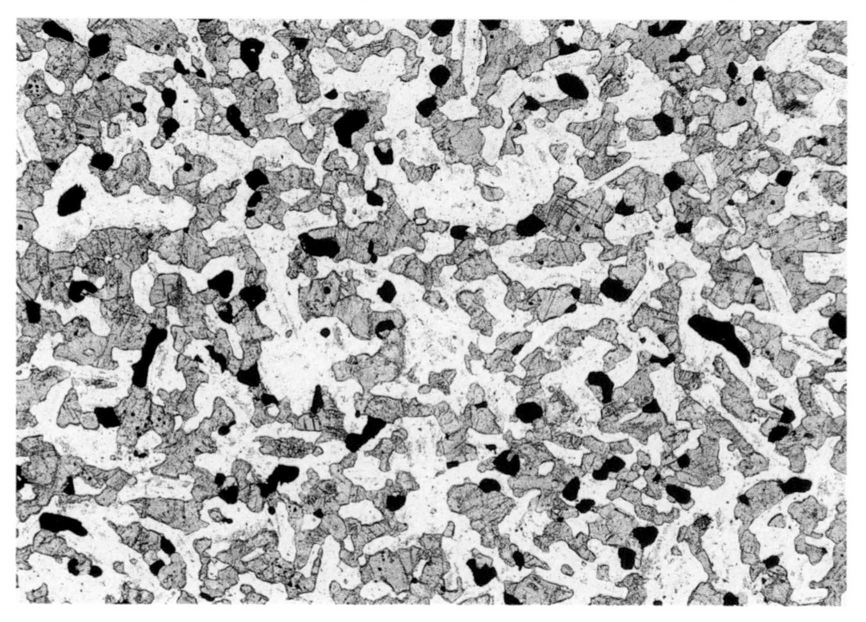

15 Fine-grained gabbro

This rock contains plagioclase, orthopyroxene, augite and magnetite; some of the orthopyroxene crystals (low birefringent mafic mineral) contain narrow lamellae of augite. Although the rock is fine grained, it is called a 'gabbro' because it is from a large intrusion; the fine grain size results from quick cooling at the intrusion margin. Another term that could be used for this rock is *microgabbro* (see p. 78).

Gabbro from chilled margin of the Skaergaard intrusion, East Greenland; magnification × 27, PPL and XPL.

Fine-grained gabbro (continued)

16 Medium-grained olivine gabbro

The spaces between the tabular crystals of plagioclase in this rock are occupied by augite and ilmenite. At the top right of the picture the plagioclase abuts onto an olivine crystal. The augite crystals contain lamellae of orthopyroxene.

Gabbro from Lower Zone b of the Skaergaard intrusion, East Greenland; magnification × 27, PPL and XPL.

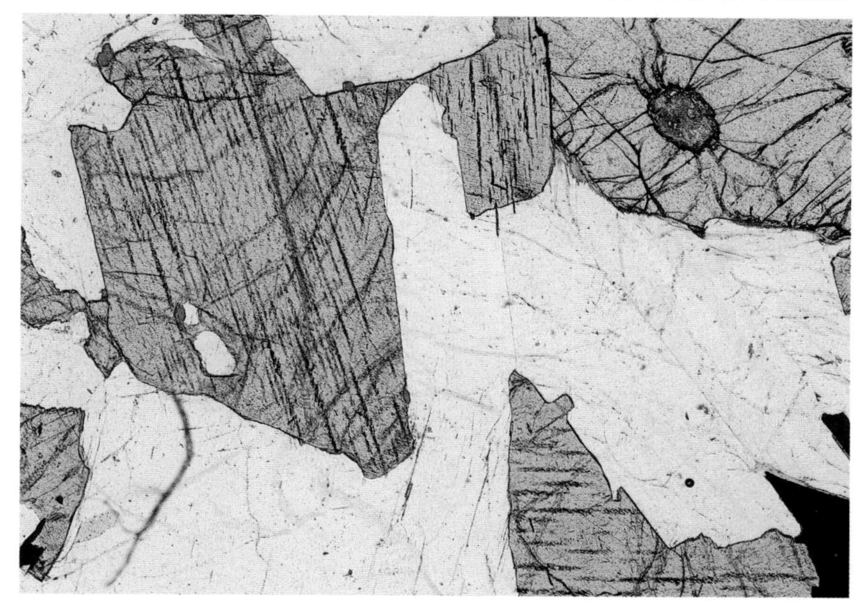

17 Coarse-grained olivine gabbro

At this magnification only parts of three large olivines and one plagioclase are visible, such that textural relations are not determinable in this single view.

Gabbro from Rhum, Scotland; magnification × *27, PPL and XPL.*

Terms indicating relative size of crystals

Equigranular – all crystals are of approximately the same size.
Inequigranular – crystals differ substantially in size. A common variety, *porphyritic* texture, involves relatively large crystals (*phenocrysts*[1]) embedded in finer-grained groundmass. (*N.B.* The same mineral may be present as both phenocrysts and groundmass.) In naming a rock with porphyritic texture the minerals present as phenocrysts should be listed and followed by the suffix -phyric, e.g. 'hornblende-pigeonite-phyric andesite'. However, if the groundmass is glassy, the term 'vitrophyre' is used, e.g. an 'olivine vitrophyre' has olivine phenocrysts set in glass; the texture in this case is referred as *vitrophyric* (**3, 142**). *Seriate* texture involves a continuous range in sizes of crystals of the principal minerals; if the crystals show a broken series of sizes, the inequigranular texture is said to be *hiatal.* Caution is necessary in the identification of seriate and hiatal textures, since the dimensions of a crystal in a thin section depend on the attitude of the intersection of the crystal in three dimensions.

[1] *The prefix* micro- *may be added to* phenocrysts *which have diameters between 0.05 and 0.5 mm (e.g. 'olivine microphenocrysts').*

18 Equigranular peridotite

Uniformly-sized olivine crystals, some of them in clots, form the bulk of this rock, with plagioclase filling the interstices. The black material is microcrystalline haematite formed by oxidation of olivines and the green material is a clay mineral.

Peridotite from the Skaergaard intrusion, East Greenland; magnification × 27, PPL and XPL.
*Additional views of equigranular rocks are shown in **43**, **113**, **117**, **125**, **130** (first photo), **134**, **140** (third photo), **168**.*

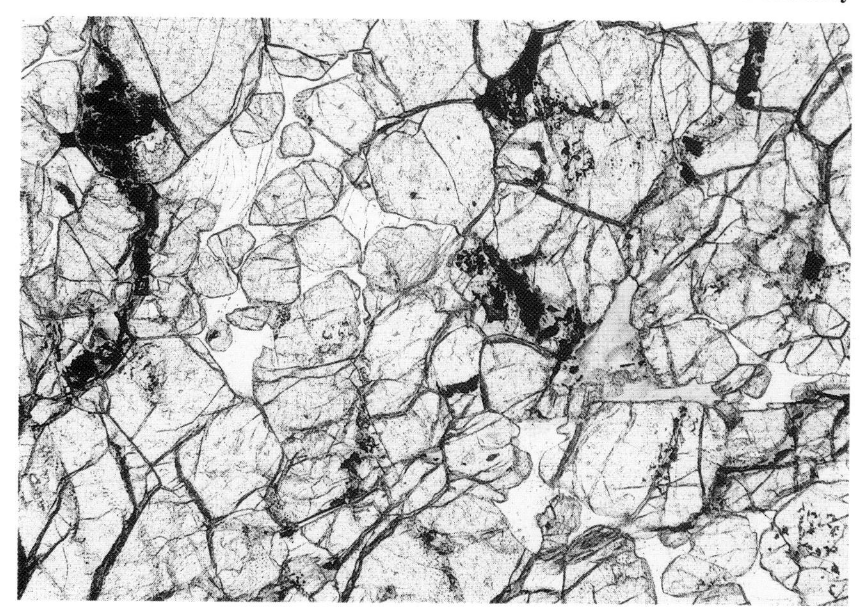

19 Porphyritic andesite

In this rock the phenocrysts (some of them in clots) of plagioclase, hornblende (khaki colour in PPL), augite (pale green in PPL) and magnetite, are surrounded by fine-grained groundmass of plagioclase, magnetite and glass.

Andesite from Siebengebirge, Germany; magnification × 23, PPL and XPL.

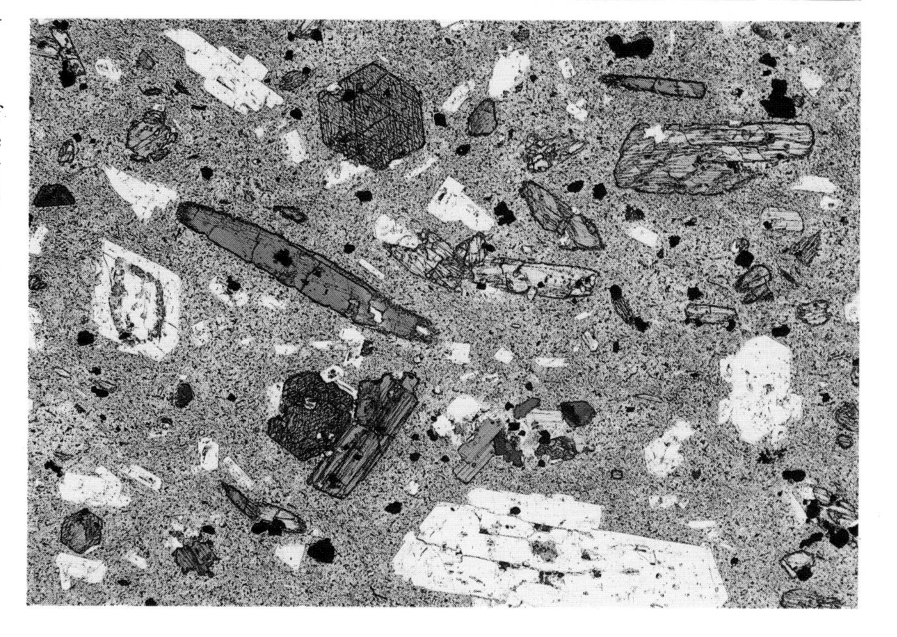

Porphyritic andesite (continued)

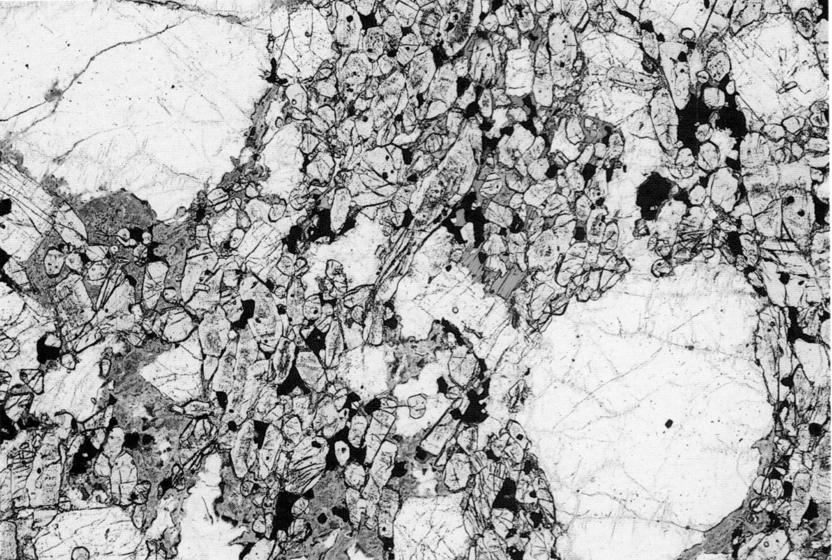

20 Leucite-phyric micro-ijolite

Two, large, shapeless crystals of leucite (very dark and showing multiple twinning in XPL photograph) are here surrounded by an equigranular groundmass consisting of crystals of elongate augite (bright interference colours), equant nepheline (grey in XPL) and interstitial biotite, leucite and magnetite. The amorphous material in the PPL view is a clay mineral.

Micro-ijolite from the Batsberg intrusion, East Greenland; magnification × 11, PPL and XPL.
Many other examples of porphyritic rocks can be seen by leafing through the book.

21 Plagioclase-augite-magnetite vitrophyre

Phenocrysts of the three minerals plagioclase, augite and magnetite, some of them in clots, are set in glass which contains crystallites of plagioclase.

Basalt from Arran, Scotland; magnification × 20, PPL.
*See **132** for another example of vitrophyre.*

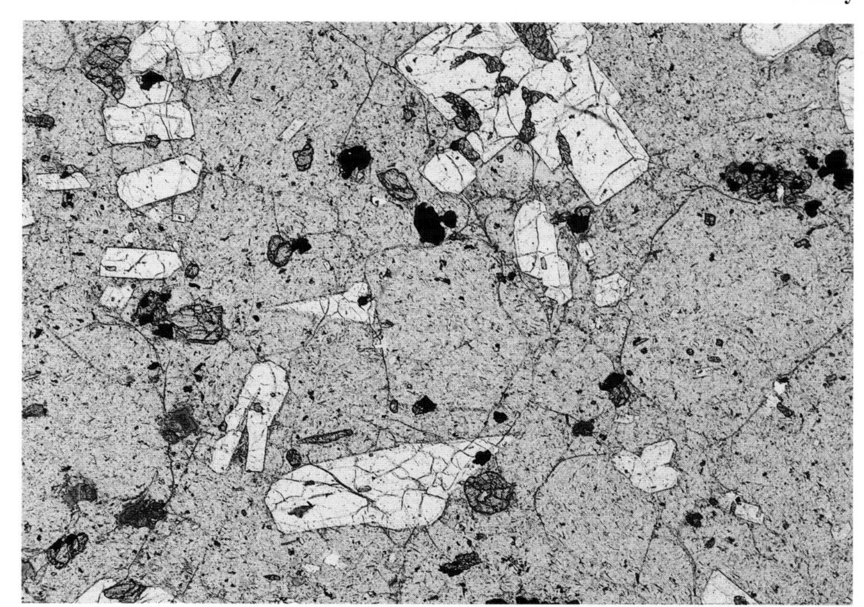

22 Seriate-textured olivine basalt

The crystals of olivine, augite and plagioclase in this basalt all show a wide range of grain size from as small as 0.01 mm up to 4 mm. Note the abundance of groundmass inclusions in some of the crystals, giving them a sponge-like appearance.

Olivine basalt from Arthur's Seat, Edinburgh, Scotland; magnification × 11, PPL and XPL.
*See **44** and **137** for other examples of this texture.*

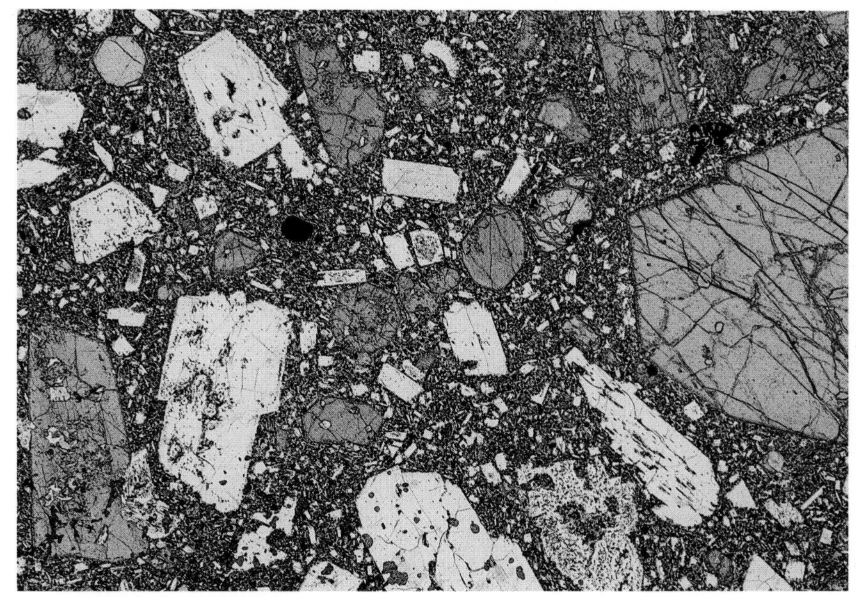

Crystal shapes

Two kinds of term are used to describe crystal shape: (1) those relating to the quality of the development of faces on crystals and (2) those specifying the three-dimensional shapes of individual crystals (p. 19).

Terms indicating the quality of the development of faces on crystals

Regrettably, three sets of words are in use to describe the same ideas, the most commonly used set being that in the first column of the following table.

Preferred terms	Synonymous terms	Synonymous terms	Meaning
Euhedral	Idiomorphic	Automorphic	Crystal completely bounded by its characteristic faces.
Subhedral	Hypidiomorphic	Hypautomorphic	Crystal bounded by only some of its characteristic faces.
Anhedral	Allotriomorphic	Xenomorphic	Crystal lacks any of its characteristic faces.

23 Euhedral olivine in olivine basalt

The photograph shows the characteristic six-sided euhedral shape of olivine in sections through the prism and dome faces. Note the slight enclosure of matrix material by one of the prism faces.

Olivine basalt from Ubekendt Ejland, West Greenland; magnification × 40, XPL.

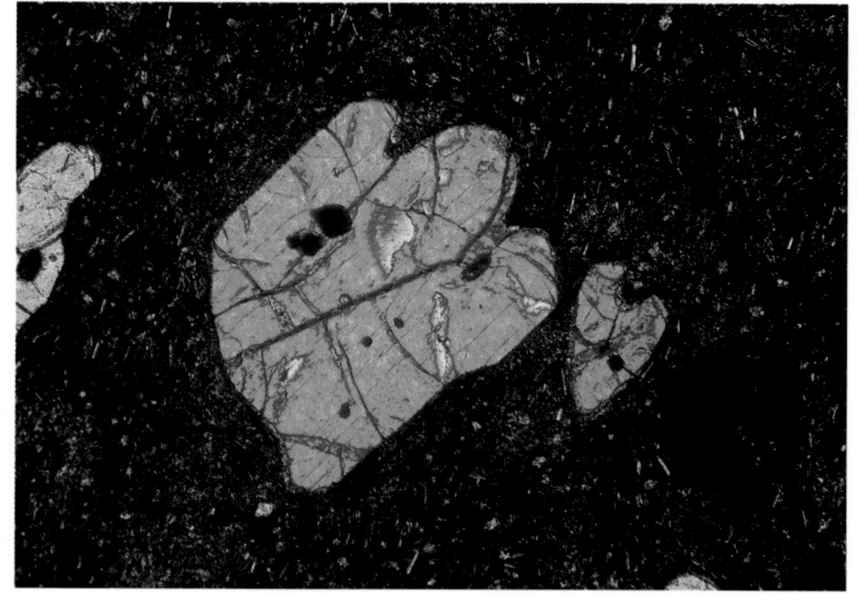

24 Subhedral olivine in picritic basalt

Some of the faces on this equidimensional olivine crystal are flat, planar ones, whereas others are curved and embayed.

Picritic basalt from Ubekendt Ejland, West Greenland; magnification × 72, XPL.

25 Anhedral olivine phenocryst in basalt

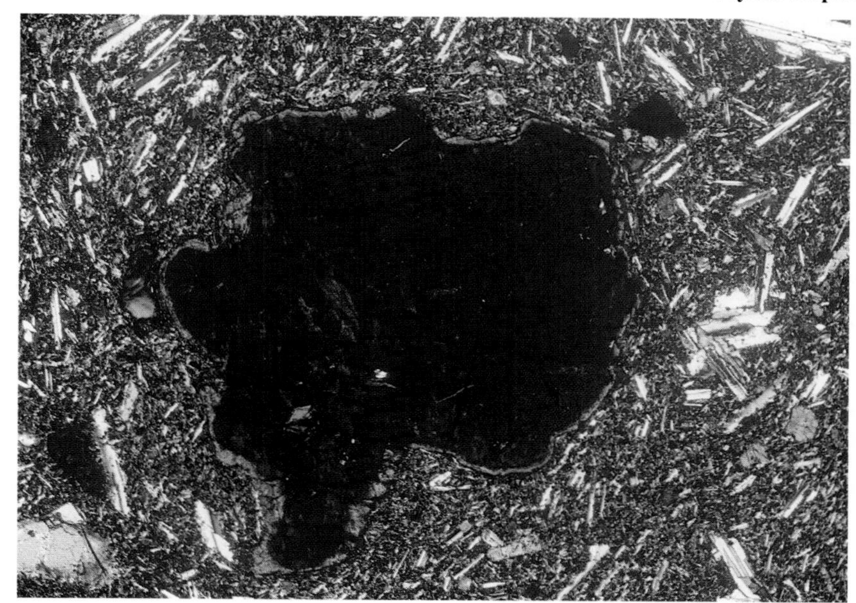

The entire perimeter of the large olivine crystal, at extinction in this picture, has an irregular outline and no planar faces are present. (The narrow brown rim on the crystal is 'iddingsite' formed by hydration and oxidation of the olivine.)

Olivine basalt from Mauritius, Indian Ocean; magnification × 32, XPL.

Terms indicating three-dimensional crystal shape

In hand specimens of coarse-grained rocks it is often possible to see the three-dimensional shape of a crystal on a broken surface. For finer-grained rocks, however, the crystals have to be examined in thin sections and the two-dimensional shapes of several crystals of different orientations used to deduce the three-dimensional shapes of the crystals in general.

General three-dimensional terms

The shape may either be an *equidimensional* (syn. *equant*) or an *inequidimensional* one, as illustrated in figs. A and B where the names applied to the various shapes are shown.

Fig. A Examples of equidimensional crystal shapes

The words *grain* and *granule* are often used for equidimensional crystals, and *drop* and *bleb* for particularly small examples.

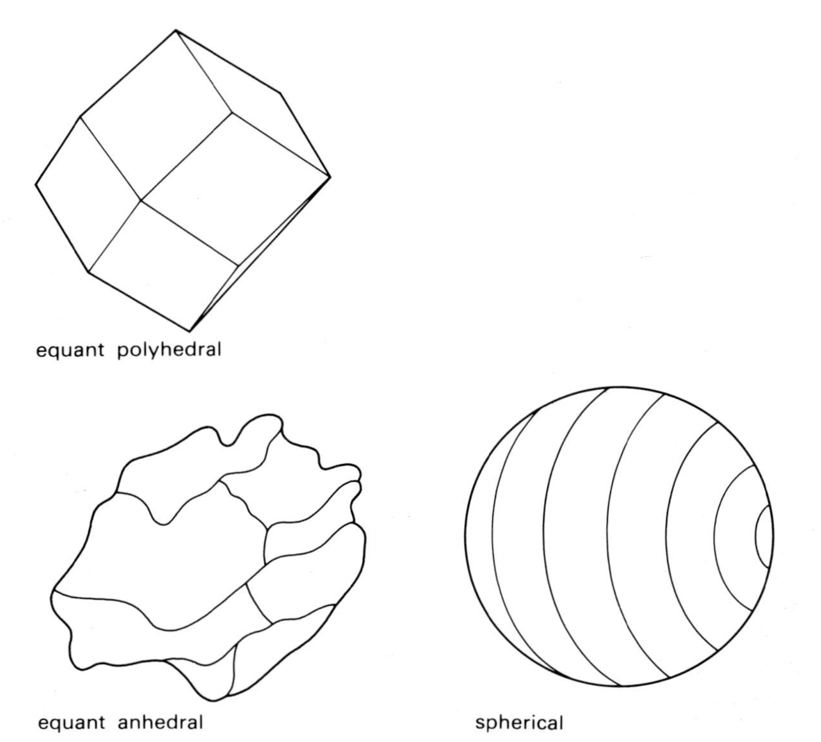

Crystal shapes

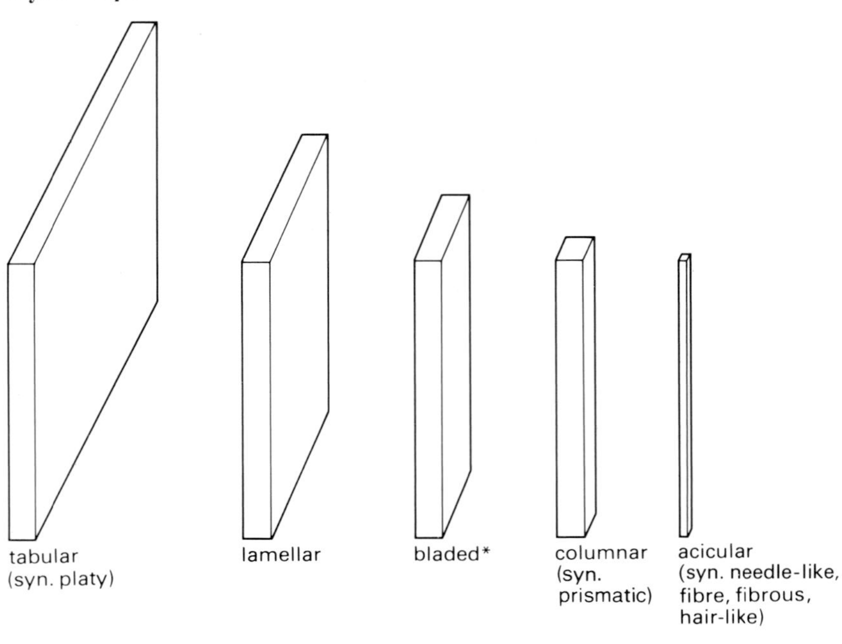

Fig. B Examples of inequidimensional shapes

N.B. Although these are euhedral examples, they could be subhedral or anhedral.

**Bladed feldspar crystals by common usage are frequently described as 'lath-shaped' or as 'laths of feldspar', in allusion to the slats (laths) in a Venetian blind.*

Specific three-dimensional terms

Skeletal, dendritic and embayed crystals

Skeletal crystals are those which have hollows and gaps, possibly regularly developed, and usually with particular crystallographic orientations. In thin section these spaces appear as embayments[1] and holes in the crystal, filled with groundmass crystals or glass. *Dendritic crystals* consist of a regular array of fibres sharing a common optical orientation (i.e. all part of a single crystal) and having a branching pattern resembling that of a tree or the veins in a leaf or a feather. In practice, many crystals can be described as either skeletal or dendritic because they have characteristics of both.

---

[1] *A common misconception among petrologists is that the terms 'embayment' and 'embayed' imply resorption of a crystal by reaction with liquid. While this may be true of some crystals (e.g.* ***29****), others (e.g.* ***26*** *and* ***27****) have embayments which probably formed during growth.*

---

26 Skeletal olivines in picritic basalt

All the large crystals in this rock are olivines and each shows a different shape in section; some are complex skeletal crystals (e.g. elongate yellow crystal on the left), others are relatively simple skeletons (e.g. equant orange crystal, middle right) and yet others have only small embayments.

Picritic basalt from Ubekendt Ejland, West Greenland; magnification × 40, XPL.

27 Skeletal olivine

While superficially resembling the euhedral outline of the olivine in **23**, the crystal occupying the bulk of this picture has a complex interior form and incomplete prism and dome faces.

Picritic basalt from Ubekendt Ejland, West Greenland; magnification × 15, PPL.

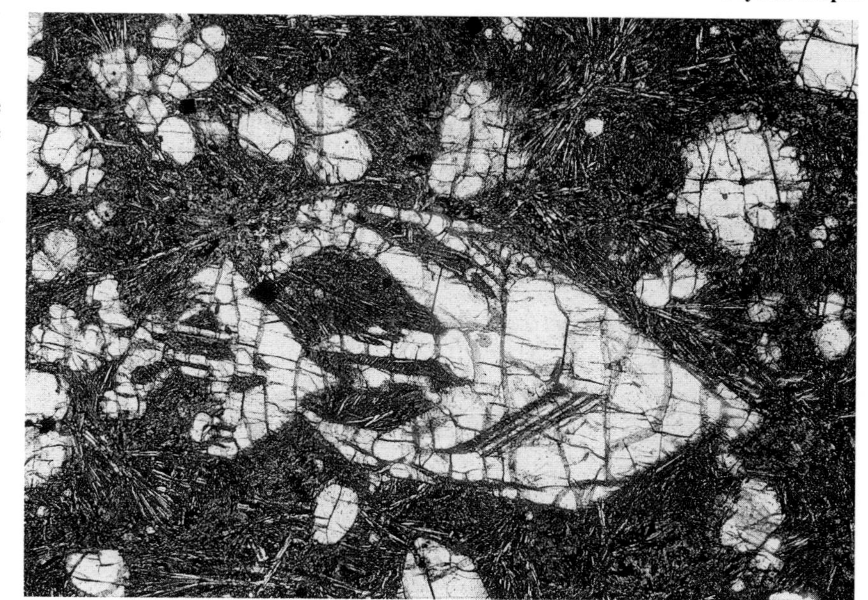

28 Dendritic olivines

All the delicate, dendritic crystals in this photograph are olivines which formed during exceedingly rapid solidification of the basalt melt, part of which became the yellow glass.

Specimen of olivine basalt melted and then cooled at 1400°/hr in the laboratory; magnification × 40, PPL.

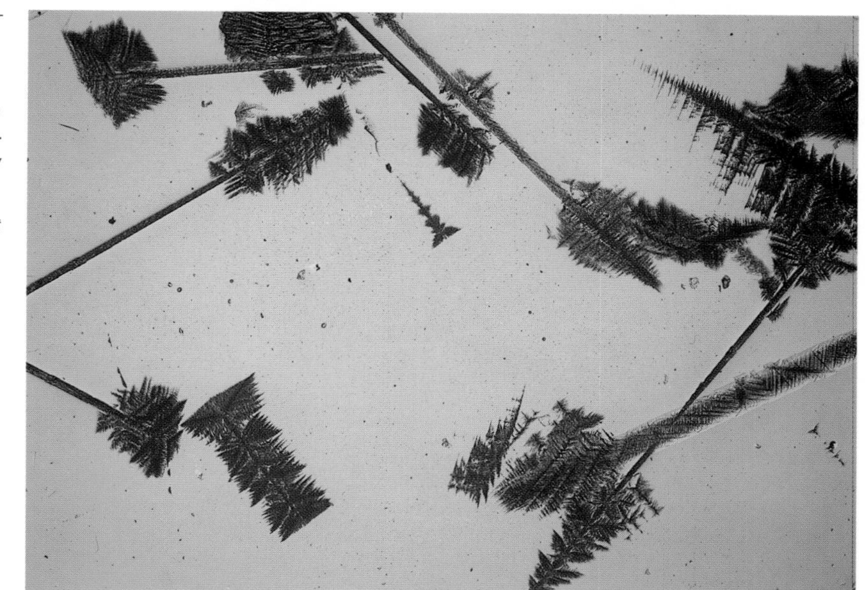

29 Embayment in augite phenocryst

The large augite crystal in this photograph contains a deep embayment filled with the basaltic groundmass. The irregular outline of this embayment distinguishes it from the embayments in the skeletal crystals in **27**. Note also the distinct marginal zoning and the delicate 'patchy zoning' within the crystal.

Olivine basalt from Arthur's Seat, Edinburgh, Scotland; magnification × 23, XPL.

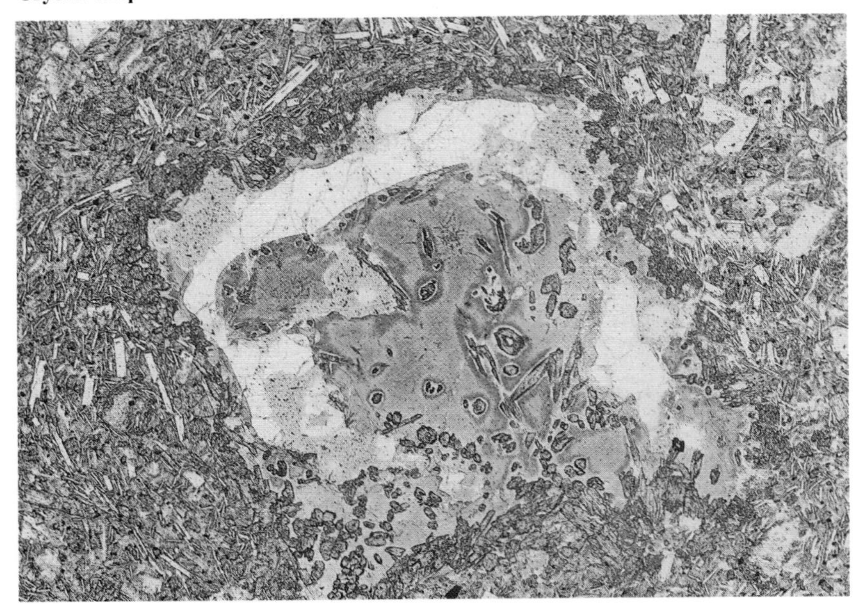

30 Embayed quartz

The deeply embayed quartz crystal in this olivine basalt contains brown glass and small, columnar, skeletal pyroxenes. It is also surrounded by a film of the glass and an aggregate of equant granular augite crystals which separate it from the basaltic groundmass.

Olivine basalt from Lassen Park, USA; magnification × 42, PPL.

Parallel-growth crystals

The term is applied to an aggregate of elongate crystals of the same mineral whose crystallographic axes are mutually parallel, or almost so. Although in thin section the individual parts of the aggregate may be isolated from one another, in the third dimension they are probably connected. A parallel-growth crystal is therefore a single, incomplete crystal formed by a particular style of skeletal growth.

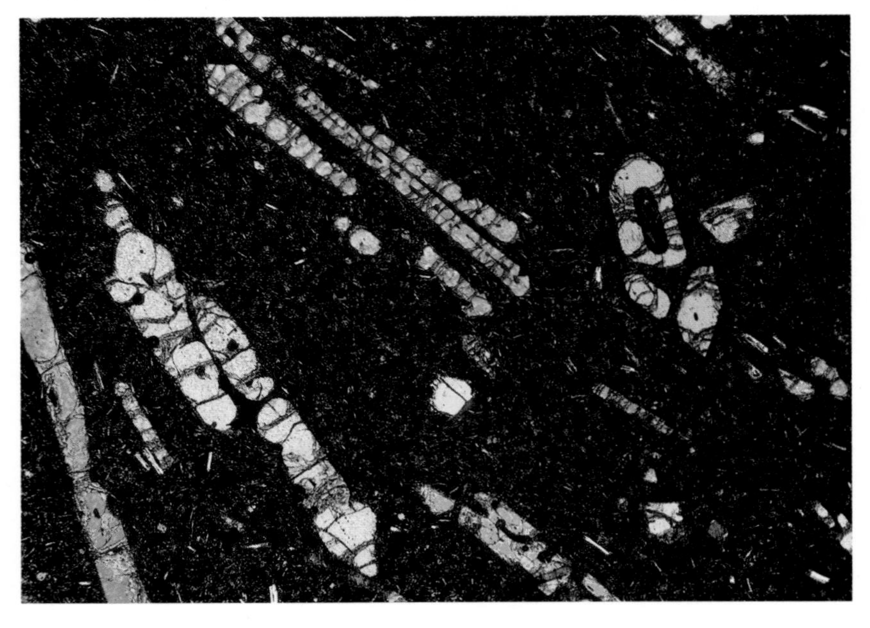

31 Olivine parallel growth

The elongate olivines near the middle of the photograph and showing blue interference colour all have the same crystallographic orientation, and hence represent a single, parallel-growth crystal. The crystal with yellowish-green interference colour shows how the parallel-growth crystal *might* appear, if sectioned at right angles.

Picritic basalt from Ubekendt Ejland, West Greenland; magnification × 23, XPL.

32 Parallel growth in a very coarse-grained rock

Here the parallel growth is of a very large olivine crystal. The actual width of the field of view is 1.7cm and this shows only a small part of the parallel growth, whose total width is 50cm and height is 150cm. The whole comprises several hundred parallel units like the ones shown here. Plagioclase and augite occupy the 'channels' between the parallel growths. In the XPL picture the polars have been rotated so that the olivine is not in extinction. The slight differences in birefringence of the olivine at the top and bottom of the picture are caused by the section being thinner there. This rock has the special textural name *harrisite*.

Feldspathic peridotite from Rhum, Scotland; magnification ×7, PPL and XPL.

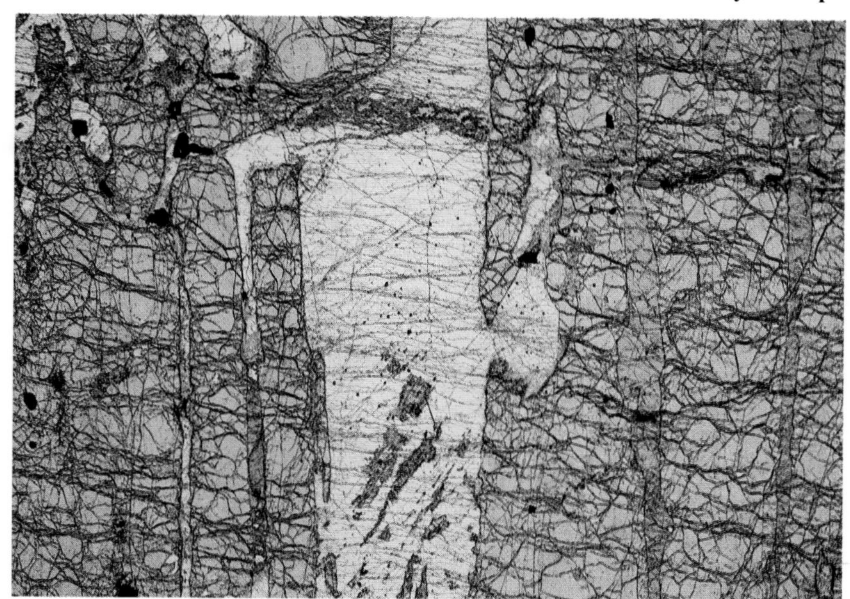

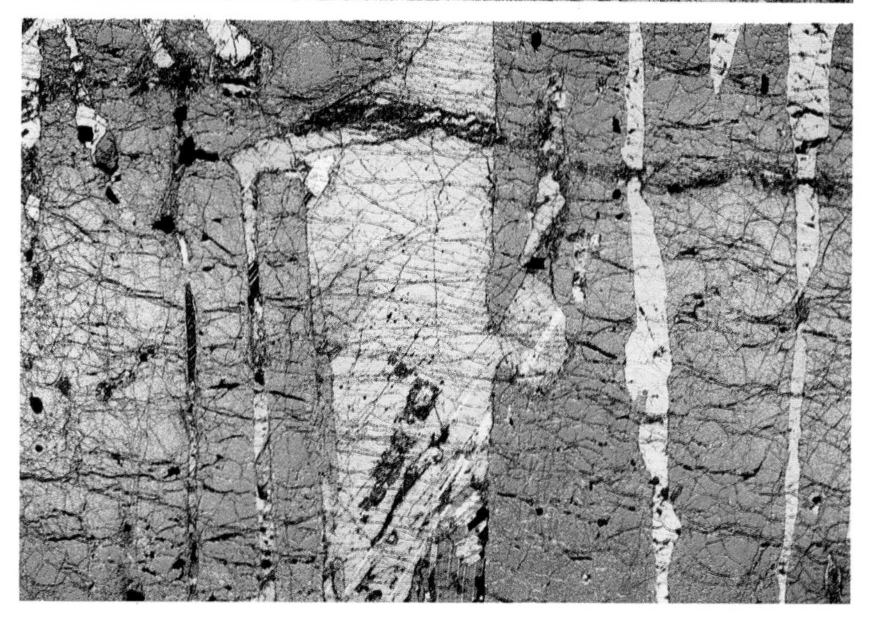

Sieve-textured crystals
These contain abundant, small, interconnected, box-shaped glass inclusions, giving the crystals a spongey, or porous, appearance.

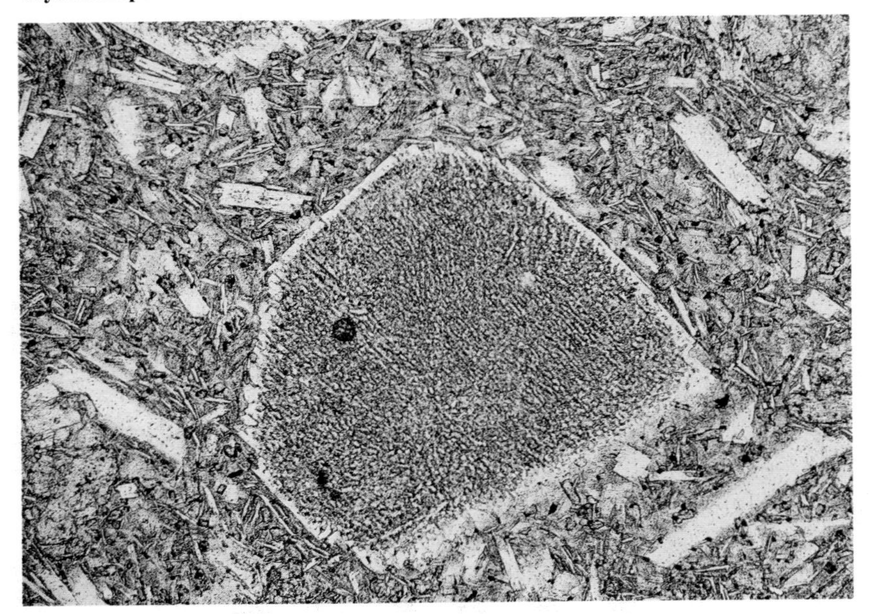

33 Sieve-textured feldspar

The core of this xenocryst consists of glass and alkali feldspar in a fine-mesh-like arrangement; the narrow rim is an overgrowth of plagioclase.

Olivine basalt from Lassen Park, USA; magnification × 62, PPL.

Elongate, curved, branching crystals
These are rarely genuinely bent, rather the curvature is caused by development of branches along the length of the crystal, each branch having a slightly different crystallographic orientation to its neighbours (e.g. **34–36**).

34 Curved branching augite

The highly coloured crystals in this photograph are complex, branching crystals of augite in subparallel alignment. They form part of a pyroxene-rich band in a differentiated dyke. (See also **71**.)

Dolerite from North Skye, Scotland; magnification × 21, XPL.

35 Branching augite in lamprophyre dyke

The acicular, aligned phenocrysts in this photograph are all of augite, forming composite, radiating, curved and branching groups. Individual needles can be seen to consist of several straight portions offset slightly from one another, and having very slightly different orientations; this gives each 'needle' its curved appearance. The margin of the dyke lay to the left. (See also **70**.)

Fourchite from Fiskaenesset area, South-west Greenland; magnification × 20, XPL.

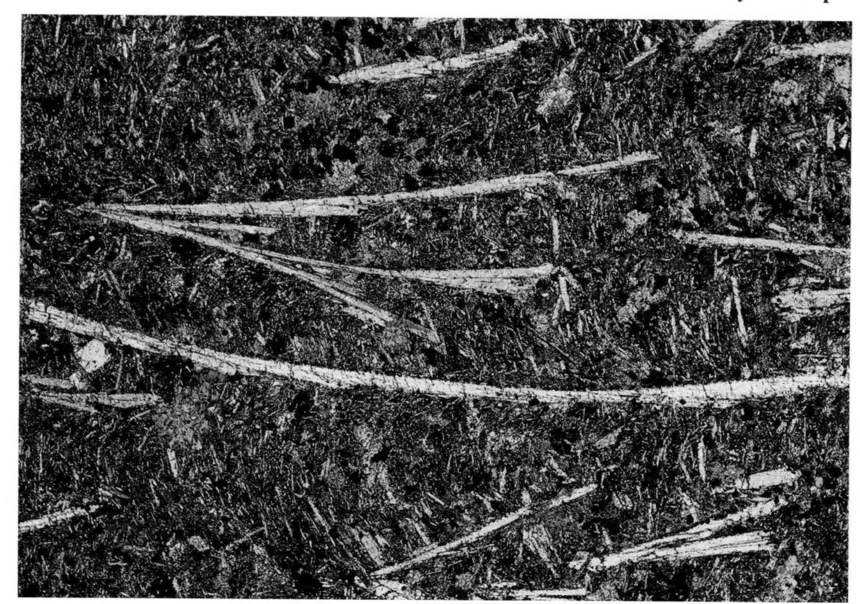

36 Curved and branching plagioclase crystals in dolerite

The large composite plagioclase crystals in this rock are elongate parallel to the *c* crystallographic axis and flattened parallel to (010). From the direction in which they branch, and from that in which the crystal at the bottom widens, it can be deduced that the crystals grew from right to left. The matrix consists of fine-grained plagioclase, olivine, pyroxene, amphibole, devitrified glass and clay minerals.

Feldspathic dolerite, Ubekendt Ejland, West Greenland; magnification × 16, PPL and XPL.

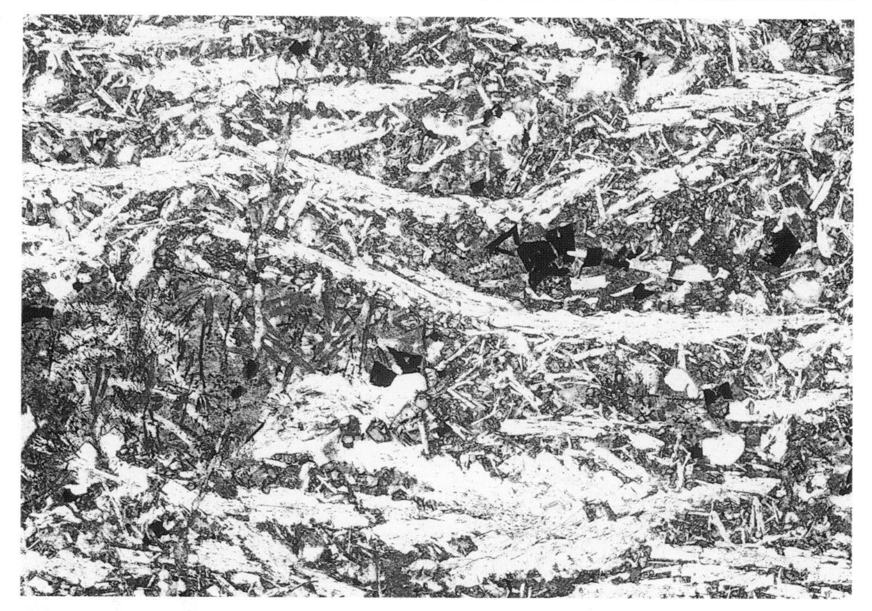

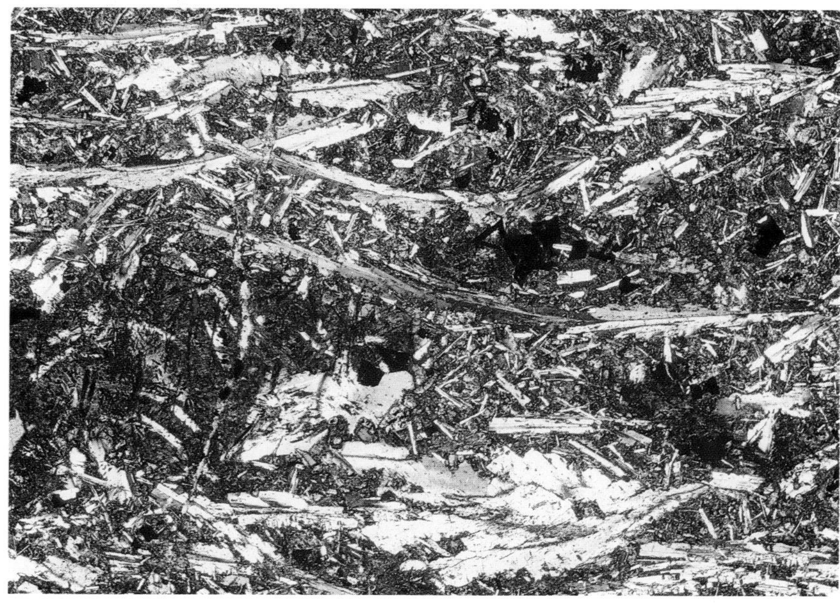

37 Composite branching augite crystal

These photographs illustrate a particularly intriguing shape of branching augite crystal: it consists of groups of slightly diverging needles, subparallel to the length of the crystal, which apparently have grown from curved branching needles oriented approximately at right angles to the crystal length. Despite the uniform interference colour of many of the needles, a sweeping style of extinction occurs when the microscope stage is rotated under crossed polars, indicating that the needles are not all of the same crystallographic orientation.

Peridotitic komatiite from Munro Township, Ontario, Canada; magnification × 52, PPL and XPL.

Pseudomorphs

It may be found that crystals in a thin section, although having the characteristic shape of a particular mineral, prove to be of another mineral, or an aggregate of crystals of another mineral. The name *pseudomorph* is used for such a crystal. If the pseudomorph has the same composition as the original crystal (e.g. 'quartz' in place of tridymite) it is known as a *paramorph*.

38 Carbonate pseudomorphs after olivine

The phenocrysts in this altered basalt show typical sections of skeletal olivine, with inclusions of groundmass in the embayments. However the photograph shows the phenocrysts to be occupied by finely crystallized carbonate, indicating that replacement of olivine has occurred.

Altered basalt from Castleton, Derbyshire, England; magnification × 27, XPL.

Another example of pseudomorphs is shown in ***149***.

Mutual relations of crystals (and amorphous materials)

The various patterns of crystal arrangement which can exist are conveniently introduced under the following headings: equigranular textures; inequigranular textures; oriented textures; intergrowth textures; radiate textures; overgrowth textures; banded textures; and cavity textures. Particular textures may belong to more than one of these categories and some also belong to the categories of **crystallinity**, **granularity** and **crystal shape**. Thus certain of the textures introduced in this section have already been mentioned and reference is made to photographs of them in previous sections.

Equigranular textures

Depending on the general shape of the crystals, three textures can be distinguished in which crystals of the principal minerals in a rock are of roughly uniform grain size:

name	synonyms	definition
euhedral granular	panidiomorphic granular	bulk of the crystals are euhedral and of uniform size
subhedral granular	hypidiomorphic granular	bulk of the crystals are subhedral and of uniform size
(anhedral)[1] granular	allotriomorphic granular (granitic and granitoid textures apply to siliceous rocks only)	bulk of the crystals are anhedral and of uniform size

Boundaries between these categories are not sharply defined and consequently the terms are applied very subjectively. Furthermore a rock may not fit neatly into a single category, thus one in which ~50% of the crystals are euhedral and ~50% anhedral might best be described as having a mixed euhedral and anhedral granular texture.

In addition to the examples of these textures in **39–43**, others may be found in **18**, **111**, **113**, **117**, **125**, **130**, **134**, **140** and **168**.

[1] *This adjective is commonly omitted from this textural name.*

39 Euhedral granular hornblendite

Rocks possessing truly euhedral granular textures are very rare. The one in this figure is a good example of a more common situation in which only some of the crystals of the principal mineral, hornblende, are euhedral and some strictly are subhedral. In contrast to **40**, there are a higher proportion of crystals with faces and the term 'euhedral granular' is therefore suggested as most appropriate. It should be appreciated, however, that another petrologist might prefer 'subhedral granular'.

Hornblendite from Ardsheal Hill, Scotland; magnification × 7, XPL.

40 Subhedral granular gabbro

The stout prismatic plagioclase feldspar crystals which dominate this rock are mostly subhedral. The anhedral interstitial crystals are of orthopyroxene, augite and magnetite.

Gabbro from Middle Zone of the Skaergaard intrusion, East Greenland; magnification × 20, XPL.

41 (Anhedral) granular troctolite

Only a few of the plagioclases in this equigranular rock possess a face and none of the olivines do. The crystals are therefore predominantly anhedral and the 'mosaic' texture is granular.

Troctolite from Garbh Bheinn intrusion, Skye, Scotland; magnification × 17, XPL.

42 Granular granite

Excepting the scarce biotite crystals, the quartz, microcline and albite crystals which make up the bulk of the rock are anhedral and have slightly interdigitating boundaries (i.e. *consertal texture* – see p. 45).

Granite from Madagascar; magnification × 13, XPL.

43 Granular lherzolite

The crystals of olivine (colourless in PPL), and pyroxenes (pale brown in PPL) which make up 95% of this rock, lack any crystal faces.

Lherzolite xenolith from the Matsoku kimberlite pipe, Lesotho; magnification × 16, PPL and XPL.

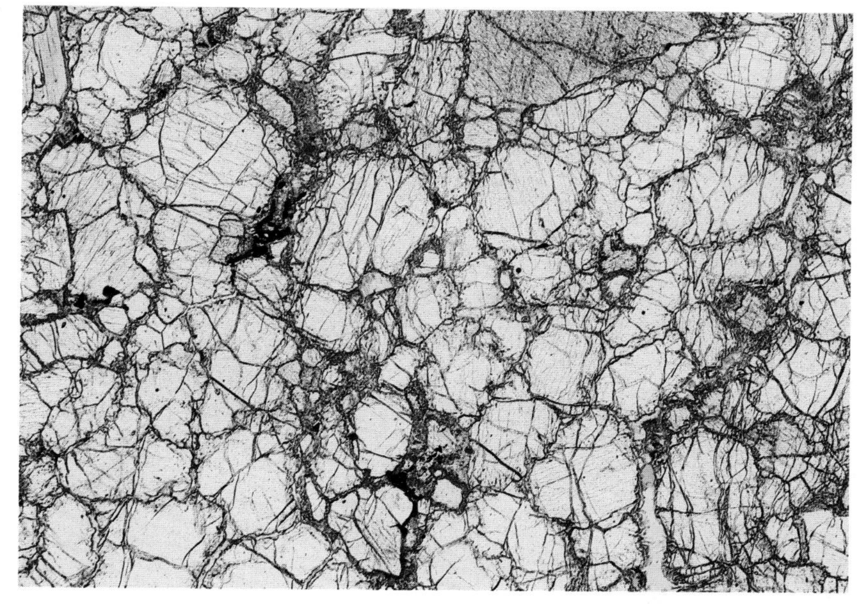

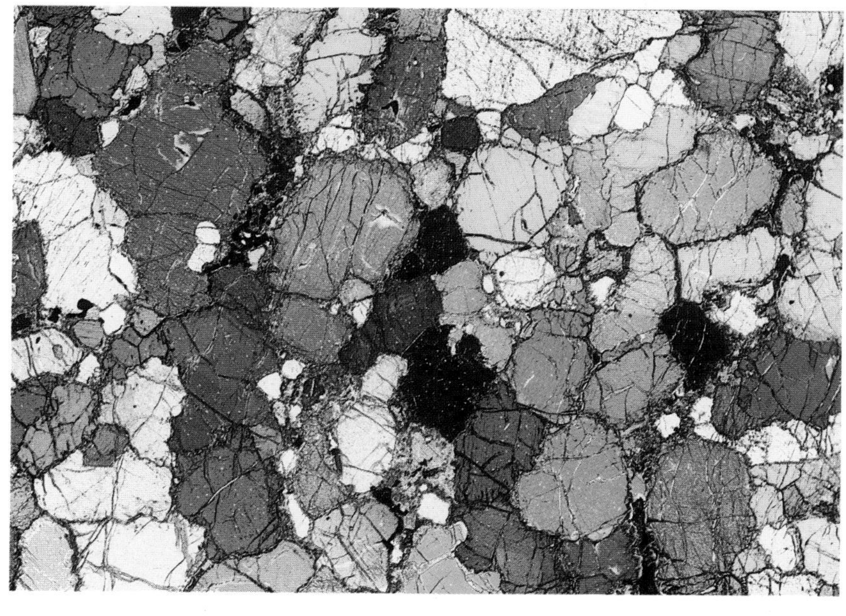

Inequigranular textures

This category includes seven kinds of texture: (a) seriate; (b) porphyritic; (c) glomeroporphyritic; (d) poikilitic; (e) ophitic; (f) subophitic; and (g) interstitial (intersertal and intergranular). It is not uncommon for a single thin section to display more than one of these textures.

Seriate texture

Crystals of the principal minerals show a continuous range of sizes. (See also p. 14.)

44 Seriate-textured basalt

This basalt, consisting of just plagioclase, augite and a small proportion of magnetite, shows a range in sizes of plagioclase and augite crystals from <0.01–0.5 mm.

Basalt from Island of Mauritius; magnification × 43, PPL and XPL.

*See **22** and **137** for other seriate-textured rocks.*

Porphyritic texture

Relatively large crystals (phenocrysts) are surrounded by finer-grained crystals of the groundmass. (See also p. 14.)

45 Augite-olivine-leucite-phyric melilitite

Augite (greyish-green and green in PPL) is present in three generations in this sample – large euhedral phenocrysts, subhedral microphenocrysts and minute groundmass crystals. The leucite occurs as colourless, equant euhedral microphenocrysts, most easily identified by their very low birefringence in the XPL picture, and the olivine as faint-grey, euhedral, columnar microphenocrysts. Note the complicated zoning pattern in one of the augite phenocrysts, the prominent marginal zoning and the line of small inclusions of groundmass crystals in another. Melilite is confined to the fine-grained granular groundmass and cannot easily be seen in these photographs.

Melilitite from Malawa, Celebes; magnification × 11, PPL and XPL.

Many more examples of porphyritic texture may be found by leafing through the book.

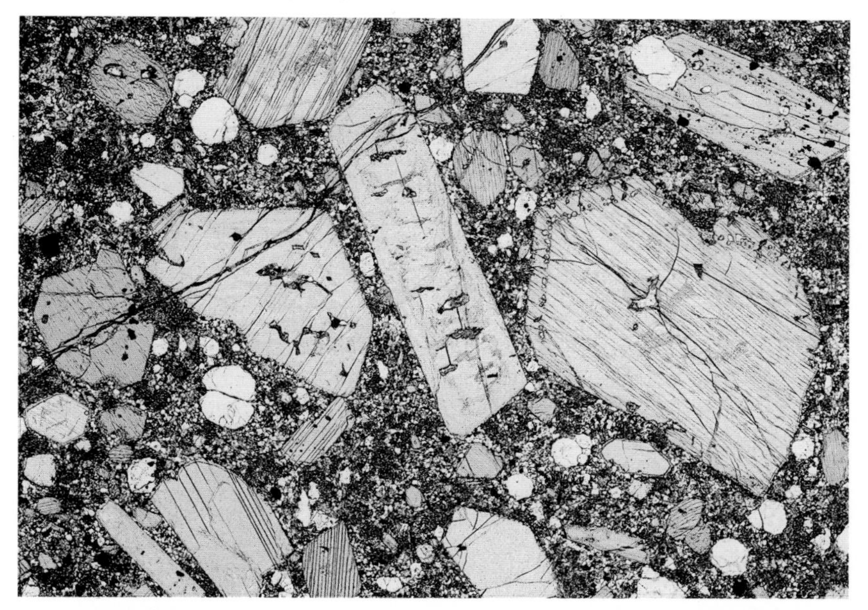

Glomeroporphyritic texture

A variety of porphyritic texture in which the phenocrysts are bunched, or clustered, in aggregates or clots called *glomerocrysts*. (A minority of petrologists maintain that the term applies only to monomineralic clots and for polymineralic clots they use the term *cumulophyric texture*.) *Glomerophyric* is usually used synonymously with *glomeroporphyritic*, though the former term strictly should be reserved for clusters of equant crystals (Johannsen, 1931). (*Synneusis texture* also describes crystal clots but includes the genetic implication that the crystals 'swam together' and is therefore best avoided.)

46 Glomeroporphyritic tholeiitic basalt

The photograph shows crystal clots of different sizes composed of plagioclase, augite and olivine crystals, enclosed by fine-grained intergranular- and intersertal textured groundmass.

Basalt from unknown locality; magnification × 11, XPL.

47 Glomeroporphyritic hawaiite

Discrete phenocrysts of plagioclase and olivine, and clots consisting of a few crystals of the same minerals, are set in a fine-grained groundmass, in places showing slight alignment of plagioclase needles. Some plagioclases in individual clots are aligned – this arrangement is common in plagioclase glomerocrysts.

Hawaiite from plateau lavas of North Skye, Scotland; magnification × 11, XPL.

*Additional views of glomeroporphyritic texture may be seen in **122**, **127**, **154** and **158**.*

Poikilitic texture

Relatively large crystals of one mineral enclose numerous smaller crystals of one, or more, other minerals which are randomly oriented and generally, but not necessarily, uniformly distributed. The host crystal is known as an *oikocryst* (or *enclosing crystal*) and the enclosed crystals as *chadacrysts*. Although *chadacrysts* are generally equant, or nearly so, they need not be uniform in size; sometimes they display progressive change in size from the interior to the margin of an oikocryst, indicating differences in extent of chadacryst growth at the time of enclosure. It is not customary to apply *poikilitic texture* to the arrangement in which scarce minute crystals of accessory minerals are embedded in a crystal, nor to that in which the enclosing mineral is approximately the same size as that included.

48 Poikilitic enclosure of olivine crystals by augite

In this photograph approximately 100 crystals of olivine of fairly uniform size are enclosed by a single augite crystal (at extinction).

Peridotite from Quarsut, West Greenland; magnification ×22, XPL.

49 Plagioclase chadacrysts enclosed by augite

Part of a single augite crystal (yellow colour), exceeding 30mm in size, is shown here enclosing plagioclase crystals, some of which form clots. The orange crystal at upper right is olivine and the crystal almost at extinction is another augite crystal.

Gabbro from North Skye, Scotland; magnification ×7, XPL.

50 Olivine gabbro containing poikilitic domains

Large plagioclases, enclosing or partially enclosing, round olivines at their margins provide a framework to this rock, the interstices of which are occupied by large augites also enclosing round olivines and small stubby crystals of plagioclase.

Olivine gabbro from Middle Border Group of the Skaergaard intrusion, East Greenland; magnification × 12, XPL.

51 Olivines enclosed by plagioclase oikocryst

Subhedral, equant olivine crystals here are enclosed in a single large plagioclase crystal.

Feldspar peridotite from Rhum, Scotland; magnification × 21, XPL.

*Additional views of poikilitic texture may be found in **111, 114** and **167**.*

Ophitic texture

This is a variant of *poikilitic texture* in which the randomly arranged chadacrysts are elongate and are wholly, or partly, enclosed by the oikocryst. The commonest occurrence is of bladed crystals of plagioclase surrounded by subequant augite crystals in dolerite (sometimes referred to as *doleritic texture*); however the texture is not confined to dolerites, nor to plagioclase and augite as the participating minerals.

Some petrologists distinguish the arrangement in which the elongate chadacrysts are completely enclosed (*poikilophitic texture*) from that in which they are partially enclosed and therefore penetrate the oikocrysts (*subophitic texture*). *Poikilophitic texture* could also be used when oikocrysts surround elongate chadacrysts of one mineral and equant chadacrysts of another.

Fine- and medium-grained rocks made up of many small oikocrysts have a patchy appearance, sometimes described as *ophimottled*.

52 Ophitic-textured alkali olivine dolerite

Two large anhedral crystals of augite enclose numerous, randomly arranged lath-shaped plagioclases. Note that many of the plagioclases are associated in groups. The larger augite crystal has a variable colour due to a chemical zoning (see p. 61).

Olivine dolerite from Shiant Isles still, Scotland; magnification × 11, XPL.

53 Subophitic texture in olivine dolerite

The photographs show plagioclase laths embedded in several augite crystals; whereas some of the plagioclases are wholly embedded, others penetrate beyond the augite crystals. The other mafic mineral present is olivine which is partially altered to a green clay-like mineral and is distinguished from the augite by its colour in the PPL view.

Olivine dolerite from unknown British source; magnification × 27, PPL and XPL.

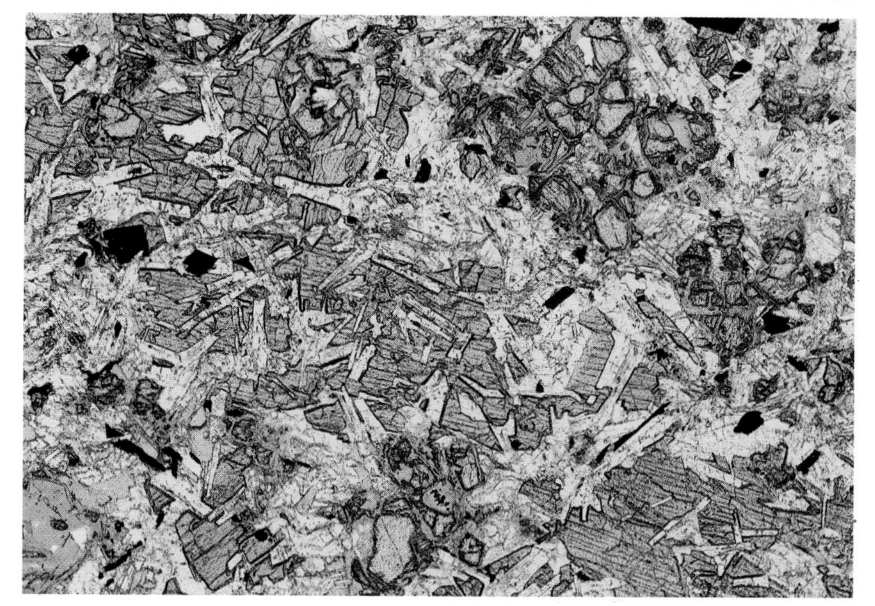

54 Subophitic alkali olivine dolerite

In this view plagioclase laths are embedded in olivine rather than pyroxene. One olivine crystal is at extinction in the XPL photograph and another shows orange interference colour. The other mafic mineral in the pictures is augite showing a purple interference colour.

Olivine dolerite from Shiant Isles sill, Scotland; magnification × 26, PPL and XPL.

*See **121**, **126**, **128** and **164** for additional examples of subophitic texture; **121** is particularly interesting because here the pyroxene is subophitically enclosed by plagioclase, and in **164** pyroxene is subophitically enclosed by kalsilite.*

55 Poikilophitic texture in olivine gabbro

For the texture shown here the term poikilophitic is preferable to ophitic because (a) the large augite encloses some equant olivines in addition to plagioclases, and (b) many of the plagioclases are not markedly elongate.

Olivine gabbro from Lower Zone a of the Skaergaard intrusion, East Greenland; magnification × 10, XPL.

56 Ophimottled texture in olivine basalt

Approximately fifty augite crystals are shown here enclosing bladed plagioclases and giving the rock a mottled or speckled appearance.

Olivine basalt from Isle of Mull, Scotland; magnification × 14, XPL.

57 Feldspar-olivine-phyric ophimottled basalt

Phenocrysts of plagioclase and olivine, some in clots, are set in fine-grained ophimottled groundmass.

Olivine basalt from Skye, Scotland; magnification × 12, XPL.

Interstitial textures

Two varieties are recognized on the basis of the material occupying the angular spaces between feldspar laths:

1. *Intersertal texture* – glass or hypocrystalline material wholly, or partly, occupies the wedge-shaped interstices between plagioclase laths. The glass may be fresh or have been altered to palagonite, chlorite, analcite or clay minerals, or it may have devitrified. If a patch of glass is sufficiently large and continuous to enclose a number of plagioclases, some petrologists would describe the texture as *hyalophitic*. (See also *hyalopilitic* texture, p. 41.)
2. *Intergranular texture* – the spaces between plagioclase laths are occupied by one, or more, grains of pyroxene (±olivine and opaque minerals). Unlike ophitic texture, adjacent interstices are not in optical continuity and hence are discrete small crystals. The feldspars may be in diverse, subradial or subparallel arrangement (see also *pilotaxitic* and *felty* textures, p. 41).

As shown by some of the photographs illustrating these textures, a single thin section may contain both types of interstitial texture in separate, but contiguous, textural domains.

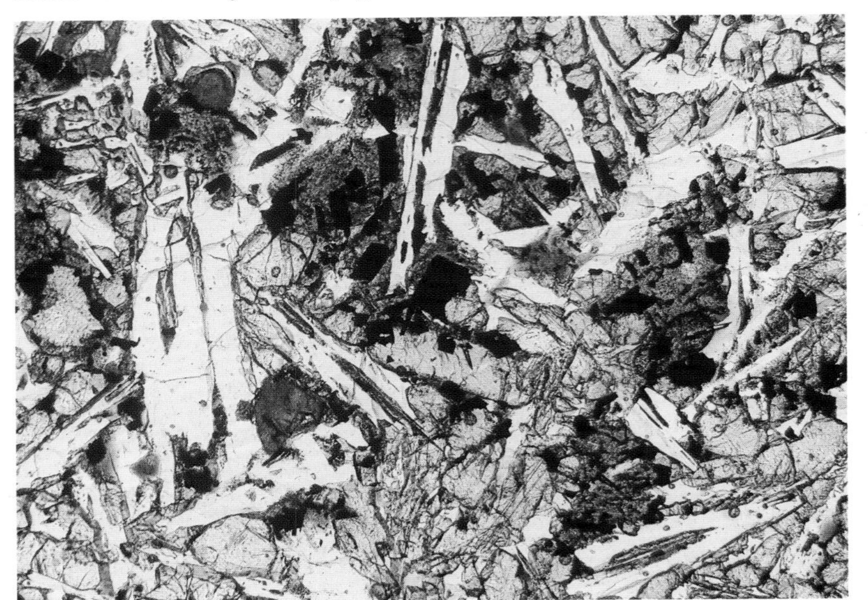

58 Intersertal (hyalophitic) texture in tholeiitic basalt

Certain parts of this photograph show lath-shaped plagioclases enclosed in pools of devitrified, deep-brown glass. Other plagioclases are surrounded by augite in a subophitic manner.

Oceanic tholeiite from Leg 34 of the Deep Sea Drilling Project; magnification × 65, PPL.

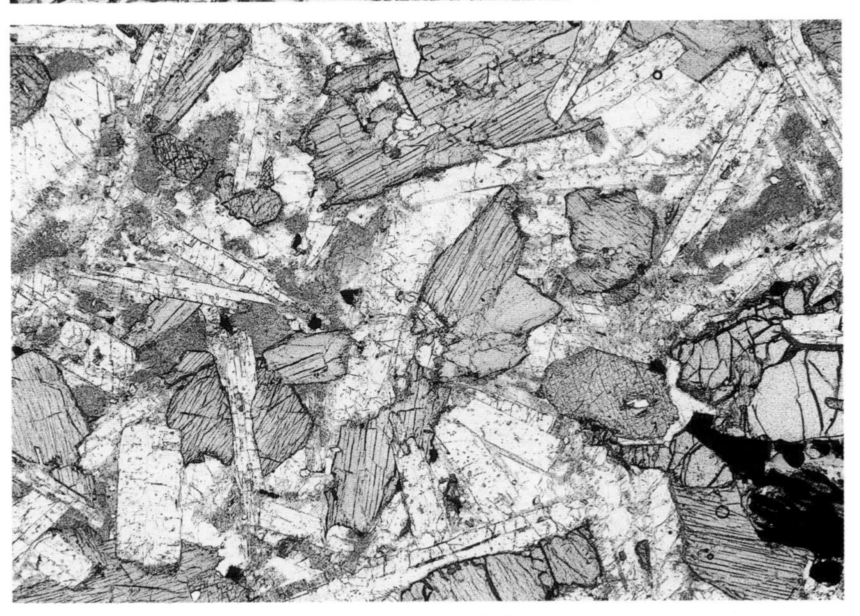

59 Intersertal texture in alkali dolerite

The intersertal texture in this dolerite consists of plagioclase crystals embedded in analcite (colourless in PPL and isotropic in XPL). Other plagioclases are partially enclosed by pyroxene in a subophitic manner. A crystal of olivine can be seen at the right-hand edge of the view in PPL.

Alkali dolerite from Howford Bridge sill, Ayrshire, Scotland; magnification × 23, PPL and XPL.

*See also **126** and **127**.*

60 Intergranular dolerite

Anhedral equant crystals of augite and pigeonite occupy the spaces between plagioclase crystals in this sample.

Dolerite from near the lower margin of Palisades sill, New Jersey, USA; magnification × 60, PPL and XPL.

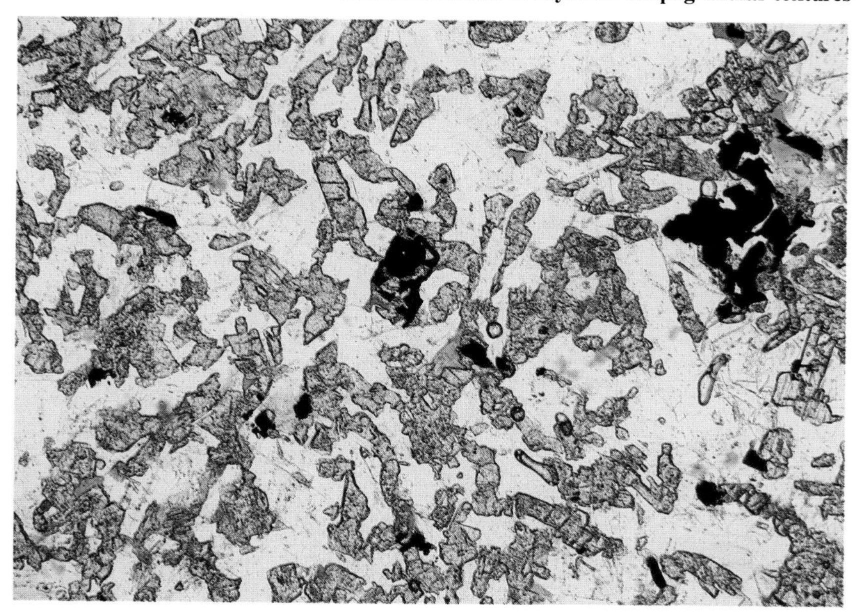

61 Intergranular olivine gabbro

In this example of intergranular texture the rock is coarse-grained and the plagioclases have a subparallel arrangement. Note that the interstitial augites are anhedral against the euhedral plagioclases.

Olivine gabbro from Lower Zone b of the Skaergaard intrusion, East Greenland; magnification × 15, XPL.

62 Tholeiitic basalt with two types of interstitial texture

In this photograph patches between some of the plagioclases are occupied by brown glass (partly devitrified) and between others by clots of small augite crystals without any glass present, i.e. domains of both intersertal and intergranular texture are present.

Tholeiitic basalt from Ubekendt Ejland, West Greenland; magnification × 27, PPL and XPL.

63 Intersertal, intergranular and subophitic textures in dolerite

All three of these textures co-exist in this rock.

Dolerite from Whin sill, Northumberland, England; magnification × 26, PPL and XPL.

Intersertal, intergranular and subophitic textures in dolerite (continued).

Oriented, aligned and directed textures

Several classes of this textural type exist: (a) trachytic texture; (b) trachytoid texture; (c) parallel-growth texture; (d) comb texture; and (e) orbicular texture.

Trachytic texture

A subparallel arrangement of microcrystalline lath-shaped feldspars in the groundmass of a holocrystalline or hypocrystalline rock.

N.B. the term is not restricted in use to rocks of trachyte composition (e.g. see groundmass of **47**).

Some petrologists subdivide trachytic texture with microlite-sized feldspars into *pilotaxitic texture* and *hyalopilitic texture*, depending on whether the material between the feldspars is crystalline or glassy.[1] Strictly, however, the microlites in these textures may be more or less aligned. (For a pilotaxitic texture in which the microlites are essentially randomly arranged the term *felty texture* exists.)

Trachytoid texture

A subparallel arrangement of tabular, bladed or prismatic crystals which are visible to the naked eye (Holmes, 1921). While the term is usually applied to crystals of feldspar, Johannsen (1931) states that it may equally well be used for oriented crystals of any other mineral.

The terms *flow* and *fluxion texture* are sometimes used as synonyms for trachytic and trachytoid textures, however they should be avoided on account of their genetic implications.

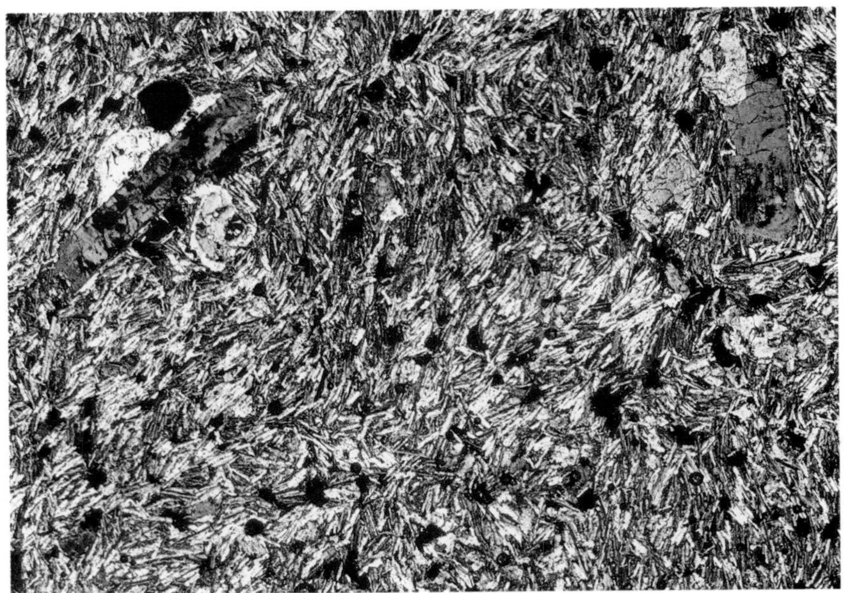

64 Trachytic texture in a trachyte

This rock illustrates trachytic texture with no glass between the small, aligned alkali feldspars (i.e. pilotaxitic variety). Note that, rather than there being a single universal alignment direction, there are several domains in the photograph, each having its own preferred direction of feldspar alignment.

Trachyte from unknown Czechoslovakian locality; magnification × 16, XPL.

65 Trachytic texture in trachyte

The somewhat stumpy groundmass alkali feldspars in this rock display a subparallel alignment which is particularly noticeable where they follow the outline of the phenocrysts.

Trachyte from unknown German locality; magnification × 15, XPL.

66 Hyalopilitic texture in rhyolitic pitchstone

The feldspar microlites in this glassy rock have a preferred elongation direction from lower left to upper right; near the feldspar phenocrysts and opaque crystals the orientation of the microlites follows the outline of these crystals. Note the tendency for the microlites to be arranged in bands.

Pitchstone from Ischia, Bay of Naples; magnification × 20, PPL.

67 Trachytoid diorite

This medium-grained rock contains aligned columnar plagioclases. The cloudy appearance to the plagioclases results from very small inclusions of iron ore and mica.

Diorite from Comrie, Scotland; magnification × 16, PPL.

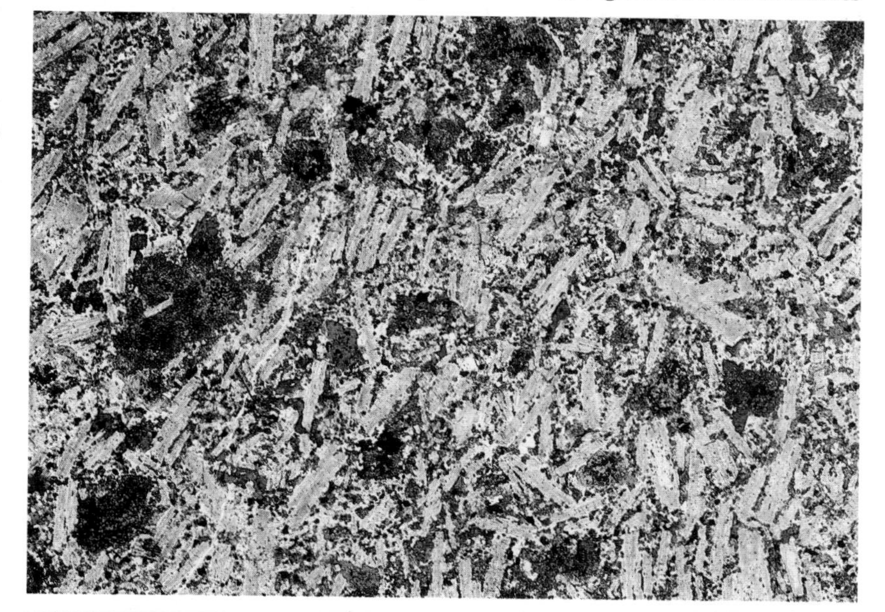

68 Trachytoid gabbro

This trachytoid texture consists of platy plagioclases, here seen edge on, stacked upon one another. Note that when this rock is sectioned parallel to the plane of the flattening, the crystal alignment would not be evident.

Gabbro from Lower Zone b of the Skaergaard intrusion, East Greenland; magnification × 12, XPL.

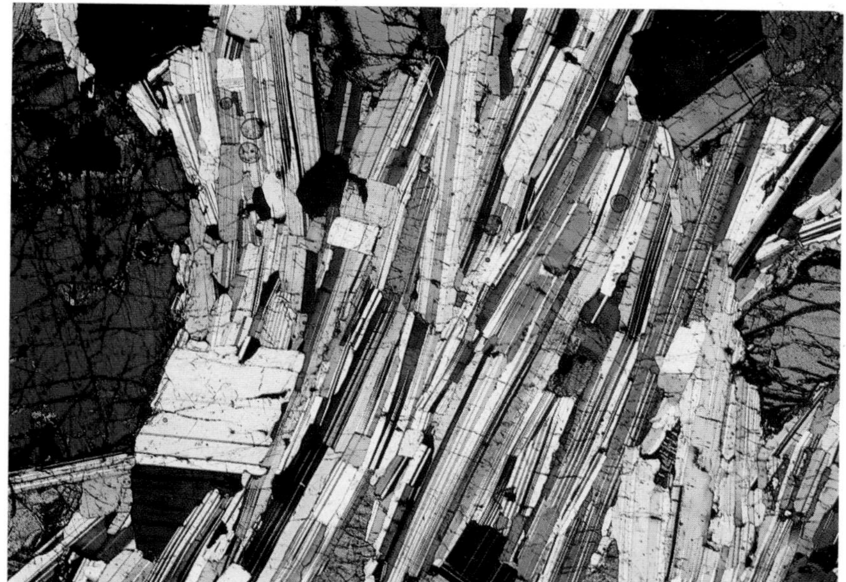

69 Olivines in trachytoid arrangement in olivine dolerite

In this view, large columnar phenocrysts of olivine, some of skeletal type, are aligned, and embedded in intergranular-textured plagioclase and augite.

Olivine dolerite from Isle of Skye, Scotland; magnification × 21, XPL.

Parallel-growth texture

A single elongate skeletal crystal which in thin section appears to consist of a clot of crystals having the same elongation direction and the same optical orientation. (For illustrations see **31** and **32**.) In rocks with trachytoid texture it is not uncommon for neighbouring parallel-growth crystals to be aligned (see **31**).

Comb texture (comb layering)

Elongate, possibly curved, branching crystals sharing the same direction of elongation. The crystals typically form a band, layer, or fringe with the elongation direction of the crystals inclined at 60–90° to the plane of the layering. (Synonyms are *Willow-Lake layering* and *crescumulate layering*, though the latter is a genetic term and, hence, should be avoided.)

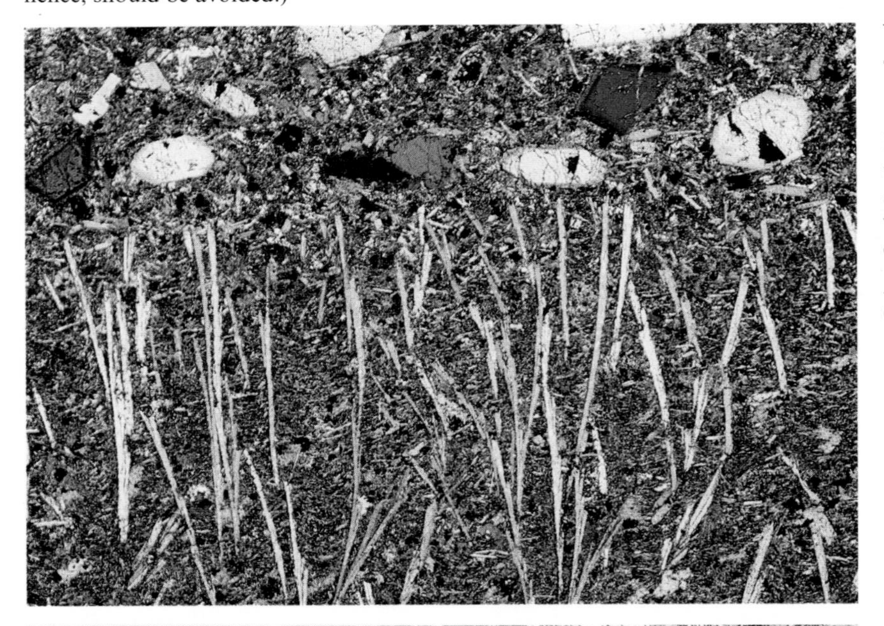

70 Pyroxene comb layer in a thin lamprophyre (fourchite) dyke

Long branching augite crystals are aligned at right angles to the boundary between the comb-layered rock (below) and pyroxene-phyric rock (above). The V of the branching widens in the direction of growth, which is away from the dyke wall. (See also **35**.)

Lamprophyre dyke from Fiskaenesset area, South-west Greenland; magnification × 8, XPL.

71 Comb layers in dolerite dyke

Two types of comb-textured layer are present in these photographs: the first and third bands from the right consist of elongate branching olivine (now largely serpentinized) and plagioclase crystals; the second and fourth bands are pyroxenite dominated by complex, elongate, branching augite crystals with scarce plagioclase crystals in between. The margin of the dyke lies to the left. (See also **34**.)

Dolerite from North-west Skye, Scotland; magnification × 8, PPL and XPL.

Comb layers in dolerite dyke (continued)

Orbicular texture (orbicular layering)

See p. 69 for definition and illustration. In connection with the group of textures being considered here, note that in some orbicules the concentric shells have elongate crystals *aligned* radially about the centre of the orbicule.

Intergrowth textures

In thin section the junction between two crystals may appear as a straight line, a simple curve, or a complex curve; in the third case the crystals interdigitate or interlock, possibly so intimately that they appear[1] to be embedded in one another. These interpenetrative patterns are all examples of *intergrowth textures*. Usually the crystals concerned are anhedral but one or both may be skeletal, dendritic or radiate. Seven varieties are distinguished here: (a) consertal texture; (b) micrographic texture; (c) granophyric texture; (d) myrmekitic texture; (e) intrafasciculate texture; (f) lamellar and blebby intergrowths; and (g) symplectite texture.

Consertal texture

The boundary between two crystals involves interdigitations and hence appears to be notched or serrated in section (Iddings, 1909; Niggli, 1954).

---

[1] *The appearance of an interdigitating boundary between two crystals, A and B, depends on the extent of interpenetration and the direction in which the boundary is sectioned: some intersections may show the crystals meeting in a complex curve; others may show crystal A enclosed in B; others may show the converse; and yet others may show each enclosing the other.*

72 Consertal texture in granodiorite

This photograph of a quartz-rich portion of the rock shows several quartz crystals with intergrown boundaries. (See also **42**.)

Granodiorite from unknown source; magnification × 43, XPL.

73 Consertal intergrowth texture in gabbro

This picture illustrates an extreme example of intergrown boundaries between crystals; the participating crystals are all augites (purple, pale yellow, grey and orange).

Gabbro from Lower Zone a of the Skaergaard intrusion, East Greenland; magnification × 25, XPL.

Micrographic texture (or graphic, if visible with the naked eye)

A regular intergrowth of two minerals producing the appearance of cuneiform, semitic or runic writing. The best-known instance is of quartz and alkali feldspar, the quartz appearing as isolated wedges and rods in the feldspar. (A micrographic intergrowth of quartz and alkali feldspar is also known as *micropegmatitic texture*.) A graphic intergrowth of pyroxene and nepheline is shown in **100**.

Granophyric texture

A variety of micrographic intergrowth of quartz and alkali feldspar which is either crudely radiate or is less regular than micrographic texture.

74 Graphic granite

Photograph of a polished hand specimen of graphic-textured granite in which the dark material is smoky quartz and the light material is alkali feldspar.

Graphic granite from unknown locality; magnification × 3.

75 Micrographic texture in aplite

Two of the crystals in this view show an intimate micrographic intergrowth of quartz and alkali feldspar. In one (middle right of XPL photograph), the alkali feldspar is at extinction, and in the other (middle left) the quartz is at extinction. (The PPL photograph is deliberately defocussed to show the Becke line in the higher-relief mineral (quartz) when the objective lens is 'raised'.)

Micro-granite from Worcester, Massachusetts, USA; magnification × 60, PPL and XPL.

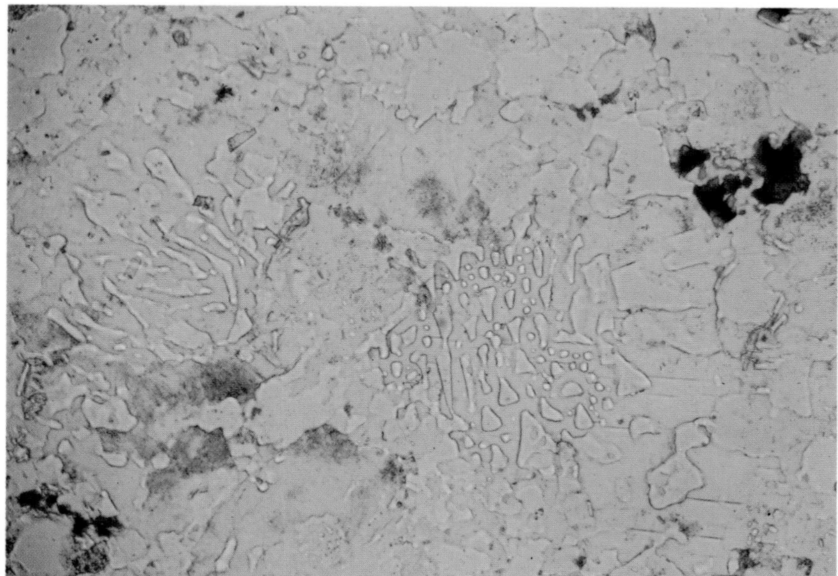

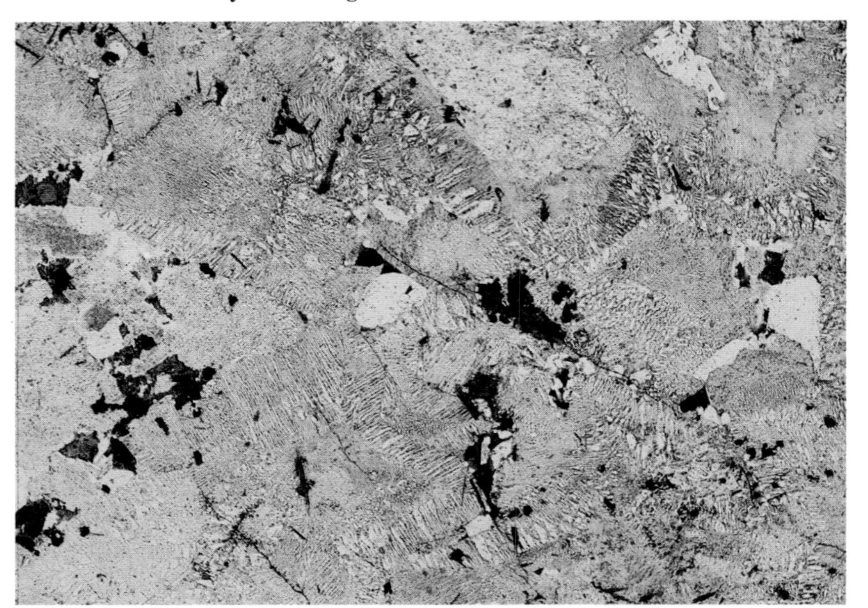

76 Micrographic and granophyric textures in microgranite

The photographs show several units of intergrown quartz and alkali feldspar; most are of micrographic type but some have a radiate arrangement (granophyric texture) at their margins. In the Scottish Hebridean igneous province, rocks like this one were formerly known as *granophyres* in allusion to their distinctive textures.

Microgranite from Eastern Red Hills of Skye, Scotland; magnification × *20, PPL and XPL.*

77 Granophyric texture

In this rock, radiate intergrowths of quartz and alkali feldspar are arranged about euhedral, equant plagioclase crystals.

Microgranite from Skaergaard intrusion, East Greenland; magnification × *37, PPL and XPL.*

Granophyric texture (continued)

Myrmekitic texture

Patches of plagioclase intergrown with vermicular quartz. The intergrowth is often wart-like in shape and is commonly to be found at the margin of a plagioclase crystal, where it penetrates an alkali feldspar crystal. The texture could be regarded as a variety of symplectite texture (see p. 53).

78 Myrmekitic texture in granite

Much of the lower part of this photograph is occupied by an intergrowth of quartz and plagioclase: this forms embayments in the microcline crystal which occupies most of the upper part of the field of view.

Granite from Rubislaw quarry, Aberdeen, Scotland; magnification × 30, XPL.

Intrafasciculate texture

Hollow, columnar plagioclase crystals filled with pyroxene.

79 Intrafasciculate texture in dolerite

This medium-grained rock has an intergrowth texture in which the gaps in the columnar plagioclase crystals are occupied by augite.

Dolerite from Garbh Bheinn intrusion, Isle of Skye, Scotland; magnification × *72, XPL.*

Lamellar and bleb-like intergrowths

Parallel lamellae, or trains of blebs, of one mineral, and all of the same optical orientation, are enclosed in a single 'host' crystal of another mineral. Well-known examples involve lamellae or blebs of sodium-rich feldspar in a host of potassium-rich feldspar (*perthitic texture*); the converse (*antiperthitic texture*); and lamellae or blebs of one pyroxene in a host of another (e.g. augite in orthopyroxene or *vice versa*, and pigeonite in augite or *vice versa*). Other examples include: ilmenite lamellae in (ulvöspinel-magnetite) solid-solution crystals; metallic iron rods, and blebs in lunar plagioclases; plagioclase lamellae in pyroxene; amphibole lamellae in pyroxene; and chrome-magnetite lamellae in olivine. Careful examination may reveal lamellae of more than one orientation and scale and sometimes even fine lamellae within coarse lamellae, i.e. multiple generations of lamellae.

Lamellar and bleb-like intergrowths are often attributed to exsolution of the lamellae and blebs from the host crystal (i.e. solid-state reaction) and the genetic term *exsolution texture* is often therefore applied to them. However, laboratory experiments in which antiperthite formed from a melt as a result of co-crystallization of two feldspars, and others in which ilmenite lamellae formed in pyroxene during co-crystallization of the two phases from the melt, highlight the danger of uncritical use of the term *exsolution texture*.

80 Microperthitic textures

Three examples of perthites are represented here.
The first photograph shows fairly broad sinuous lamellae of albite traversing the tartan twinning of a microcline crystal.

Specimen from pegmatite, Topsham, Maine, USA; magnification × 16, XPL.

The second photograph shows narrow albite lamellae forming a braided pattern in an orthoclase host (upper centre).

Specimen from granite, Ratagan, Scotland; magnification × 34, XPL.

The third photograph shows two large areas of the field of view with different orientations of crystals consisting of an intimate intergrowth of a potassium-rich feldspar and a sodium-rich feldspar. In each case the darker grey colour represents the potassium-rich feldspar. The proportions of the two materials are approximately equal so that neither is clearly the host – in this case the feldspar intergrowth is known as a *mesoperthite*.

Specimen of nepheline syenite from Langesund fjord, Norway; magnification × 32, XPL.

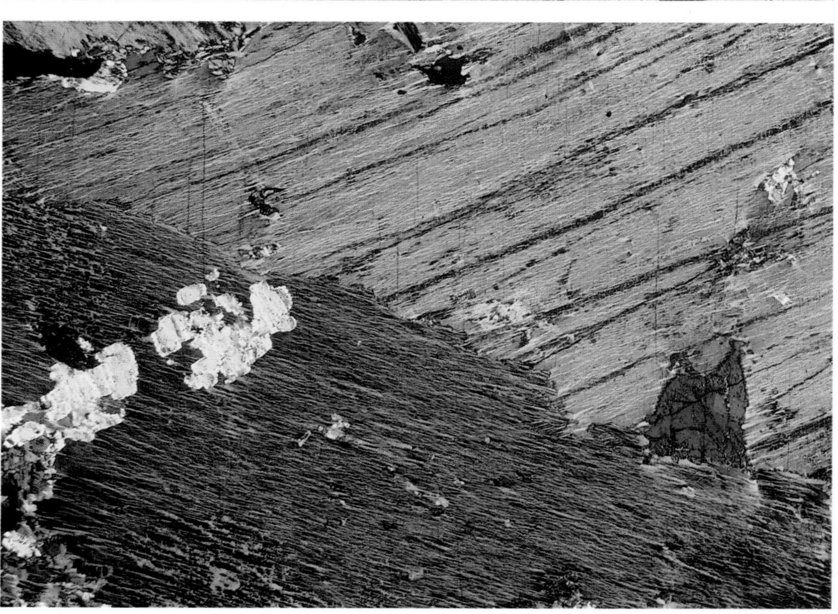

81 Antiperthitic texture in tonalitic gneiss

The poorly aligned, bleb-like inclusions in the plagioclases in this rock are potassium-rich feldspar of intermediate structural state (i.e. orthoclase). It is likely that the texture formed in this rock during prolonged high-grade regional metamorphism rather than during crystallization of magma.

Tonalitic gneiss from Scourie, North-west Scotland; magnification × 20, XPL.

82 Lamellar intergrowths of two pyroxenes in gabbro

The host crystal to the lamellae is an orthopyroxene (close to extinction); it contains two kinds of lamellae – relatively broad and continuous ones of augite, and narrower discontinuous ones of augite, inclined to the broad variety.

Gabbro from Bushveld intrusion, South Africa; magnification × 9, XPL.

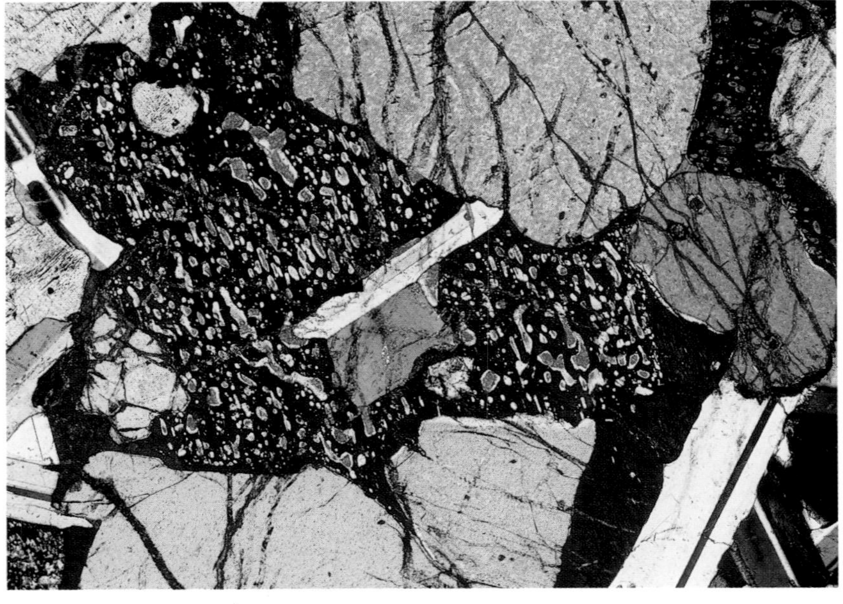

83 Bleb-like intergrowth of augite in orthopyroxene in olivine gabbro

In this sample blebs of augite are embedded in an orthopyroxene host, forming an 'emulsion-like' texture. Though the blebs are irregular in shape they have a common elongation direction and the same optical orientation.

Olivine gabbro from Lower Zone b of the Skaergaard intrusion, East Greenland; magnification × 27, XPL.

Symplectite texture

An intimate intergrowth of two minerals in which one mineral has a vermicular (wormlike) habit.

84 Symplectite of iron ore and orthopyroxene

Iron ore (probably ilmenite) and a small crystal of orthopyroxene are intimately intergrown in a vermicular fashion in the spaces between plagioclase, augite and ilmenite crystals.

Olivine gabbro from Lower Zone b of the Skaergaard intrusion, East Greenland; magnification × 72, PPL.

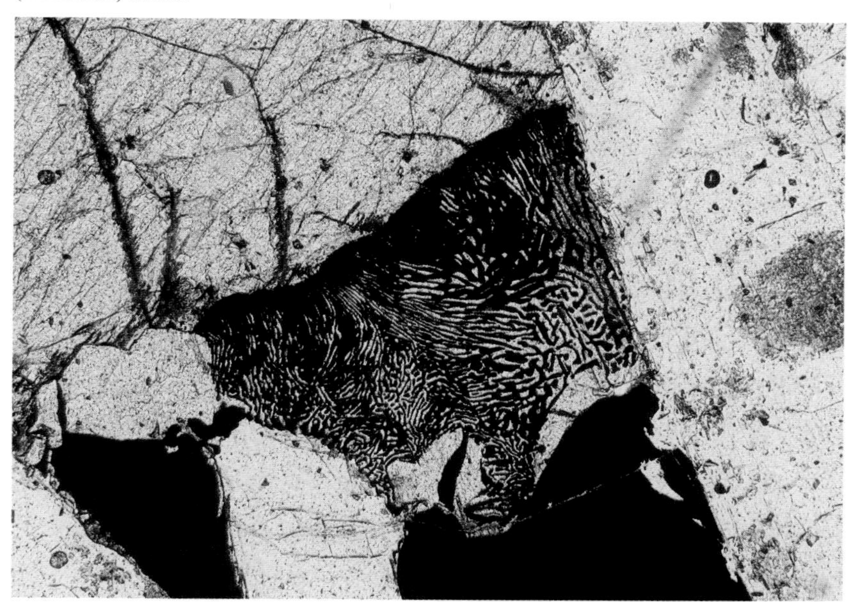

85 Fayalite-quartz symplectite

Between the opaque mineral (ilmenite) and the silicate minerals in this rock, there exists a complex boundary consisting of a narrow rim of fayalite immediately adjacent to the opaque mineral, which in places abuts onto a symplectite intergrowth of fayalite and quartz. The fayalite in the intergrowth and that which rims the ilmenite have the same optical orientation.

Ferrogabbro from Upper Zone b of Skaergaard intrusion, East Greenland; magnification × 32, PPL and XPL.

Fayalite-quartz symplectite (continued)

Radiate textures

Radiate textures are those in which elongate crystals diverge from a common nucleus. They are most frequently found in fine-grained rocks, but not exclusively; for example, **34**, **35**, **36**, **70** and **71** show large branching pyroxene, plagioclase and olivine crystals in fan-shaped radiate arrangements. A remarkably large number of terms exists to describe the various patterns, including: fan, plume, spray, bow-tie, spherical, sheaf-like, radiate, radial, axiolitic, spherulitic and variolitic. All except the last three (which are defined and illustrated here), are of self-evident meaning.

Spherulitic texture

Spherulites are approximately spheroidal bodies in a rock: they are composed of an aggregate of fibrous crystals of one or more minerals radiating from a nucleus, with glass or crystals in between. The acicular crystals may be either single, simple fibres or each may have branches along its length; any branches may or may not share the same optical orientations as their parents. The most common occurrence of spherulitic texture is a radiate aggregate of acicular alkali feldspars with glass between them, though quartz or other minerals may be present, resulting in an intergrowth texture. Should the spherulite have a hollow centre it is known as a *hollow spherulite*, and if it comprises a series of concentric, partially hollow shells, the term *lithophysa* is used.

Axiolites differ from spherulites in that radiating fibres extend from either end of a linear nucleus (i.e. from a small acicular crystal) rather than a point. They could be regarded as a variety of overgrowth texture (p. 58), as indeed could those spherulites which grow about visible crystals rather than on submicroscopic nuclei (e.g. **88**).

86 Plagioclase spherulite in dolerite

This spherulite comprises approximately twenty elongate crystals of plagioclase, each having a different optical orientation. It is an 'open' spherulite, in the sense that there is much space between individual plagioclase crystals; the spaces are occupied by coarse augite, columnar plagioclases not related to the spherulite, and smaller spherulites.

Dolerite from Garbh Bheinn intrusion, Skye, Scotland; magnification × 32, PPL and XPL.

*See **126** for a similar example.*

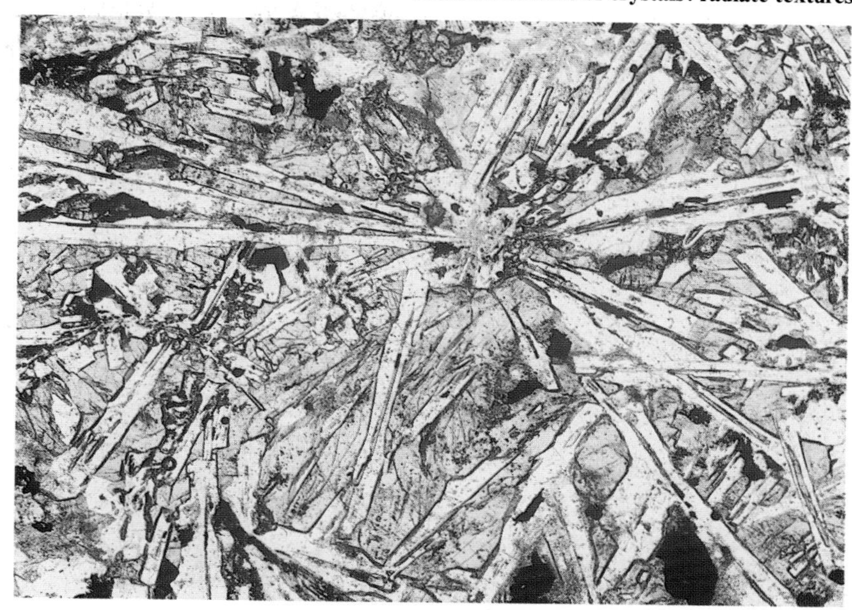

87 Spherulite in rhyolite

The spherulite at the centre of this photograph consists of a dense mass of very fine intergrown needles of both quartz and alkali feldspar radiating from a common nucleus. Above and below, the spherulite abuts onto others, whereas to left and right there is glass.

Rhyolite from Hlinik, Hungary; magnification × 27, XPL.

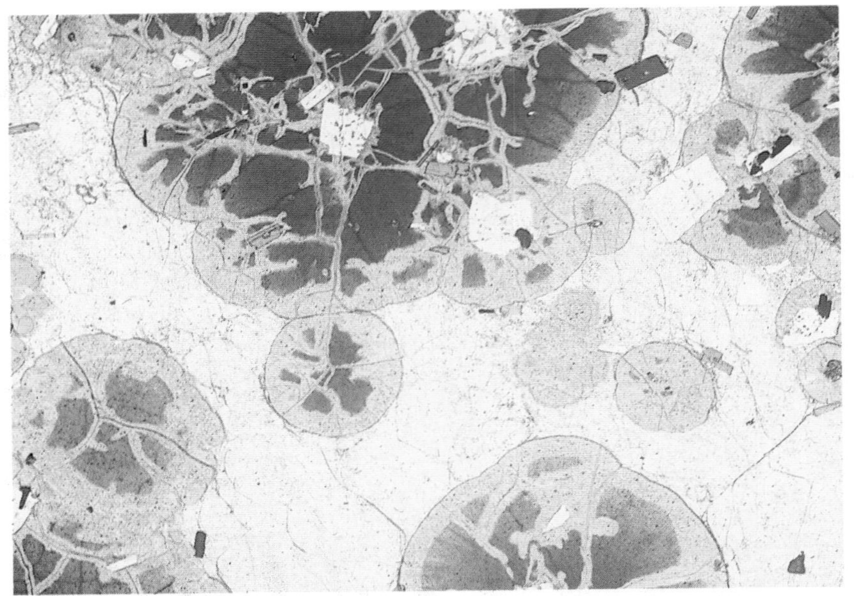

88 Compound spherulites in rhyolite

Both single and compound, or clumped, spherulites are surrounded by glass in this photograph. The spherulites enclose microphenocrysts of plagioclase and biotite. The colour variation in the spherulites is caused by variations in density of fibres.

Rhyolite from Glashutte, Hungary; magnification × 12, PPL and XPL.

Variolitic texture

A fan-like arrangement of divergent, often branching, fibres; usually the fibres are plagioclase and the space between is occupied by glass or granules of pyroxene, olivine or iron ore. This texture differs from spherulitic in that no discrete spherical bodies are identifiable; in fact, each fan as seen in thin section is a slice through a conical bundle of acicular crystals.

89 Variolitic olivine dolerite

The olivine phenocrysts in this sample are set in a groundmass consisting of many fans of diverging plagioclase needles with augite crystals in the interstices. Note how all the fans diverge from lower right to upper left, indicating progressive solidification in this direction. Note also the branching character of some of the plagioclase fibres.

Olivine dolerite from Skye, Scotland; magnification × 27, PPL and XPL.

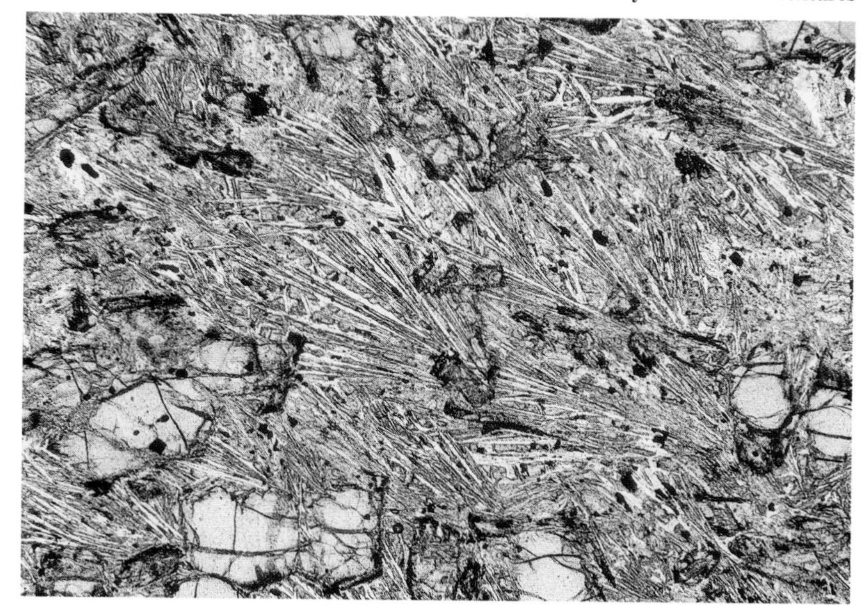

90 Radiate intergrowth of plagioclase and augite in dolerite

This unusual radiate texture occupying the centre of the view consists of two, mutually perpendicular, columnar, plagioclase crystals, the elongate gaps in which have a radiate distribution; these gaps are occupied by a *single* augite crystal, rather than by many crystals. This kind of radiate texture differs from a spherulite; it is more akin to skeletal growth (p. 20).

Dolerite from Ingia intrusion, West Greenland; magnification × 27, PPL and XPL.

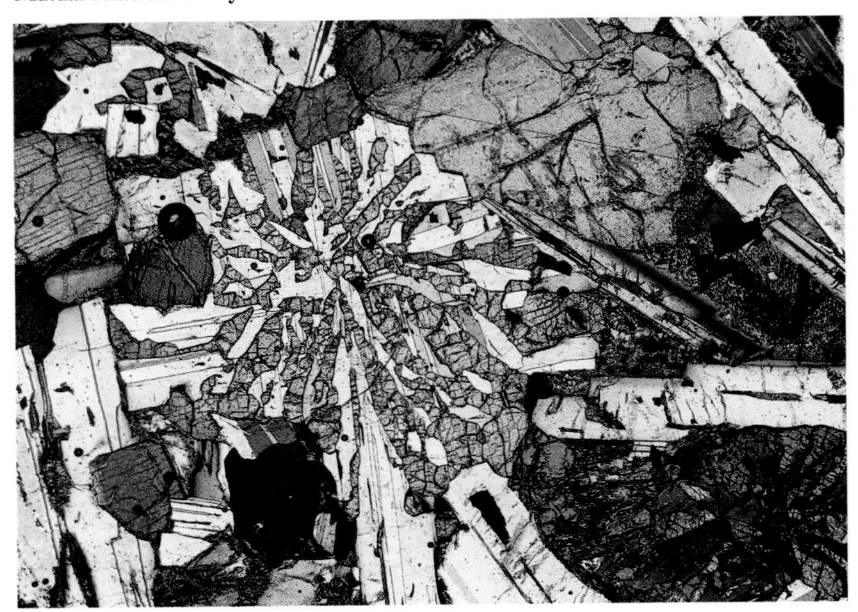

Radiate intergrowth of plagioclase and augite in dolerite (continued)

Overgrowth textures

This term applies to textures in which a single crystal has been overgrown either by material of the same composition, or by material of the same mineral species but different solid-solution composition, or by an unrelated mineral. There are three types: (a) skeletal and dendritic overgrowths; (b) corona texture; and (c) crystal zoning.

Skeletal or dendritic overgrowths

Porphyritic rocks with a glassy or very fine-grained groundmass may show delicate fibres or plates extending from the corners or edges of the phenocrysts. The overgrowth and the phenocryst need not be the same mineral.

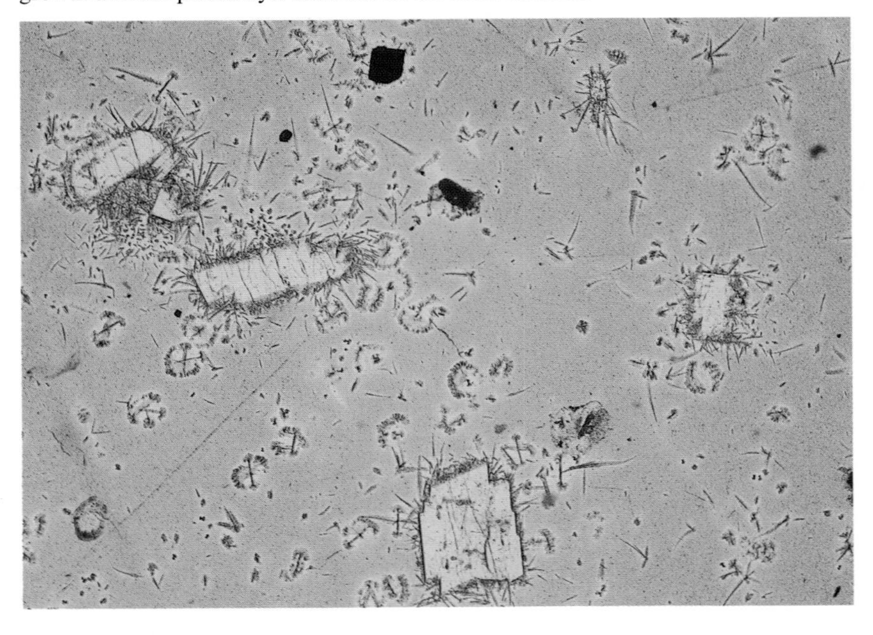

91 Overgrowth textures in rhyolitic pitchstone

The faces of the phenocrysts of alkali feldspar and magnetite in this glassy rock have acted as locations for nucleation of dendritic overgrowths of (?) alkali feldspar. Dendritic crystallites are also present in the glassy groundmass.

Pitchstone from Arran, Scotland; magnification × 31, PPL.

Corona texture

A crystal of one mineral is surrounded by a rim, or 'mantle', of one or more crystals of another mineral, e.g. olivine surrounded by orthopyroxene, or biotite surrounding hornblende. Such relationships are often presumed to result from incomplete reaction of the inner mineral with melt or fluid to produce the outer one and for this reason the equivalent genetic terms *reaction rim* and *reaction corona* are frequently used. The special term *Rapakivi texture* is used to describe an overgrowth by sodic plagioclase on large, usually round, potassium-feldspar crystals, and *kelyphitic texture* is used for a microcrystalline overgrowth of fibrous pyroxene or hornblende on olivine or garnet.

92 Corona texture

In the centre of the photographs a twinned and zoned augite crystal is mantled by green hornblende of non-uniform width.

Quartz diorite from Mull of Galloway, Scotland; magnification × 43, PPL and XPL.

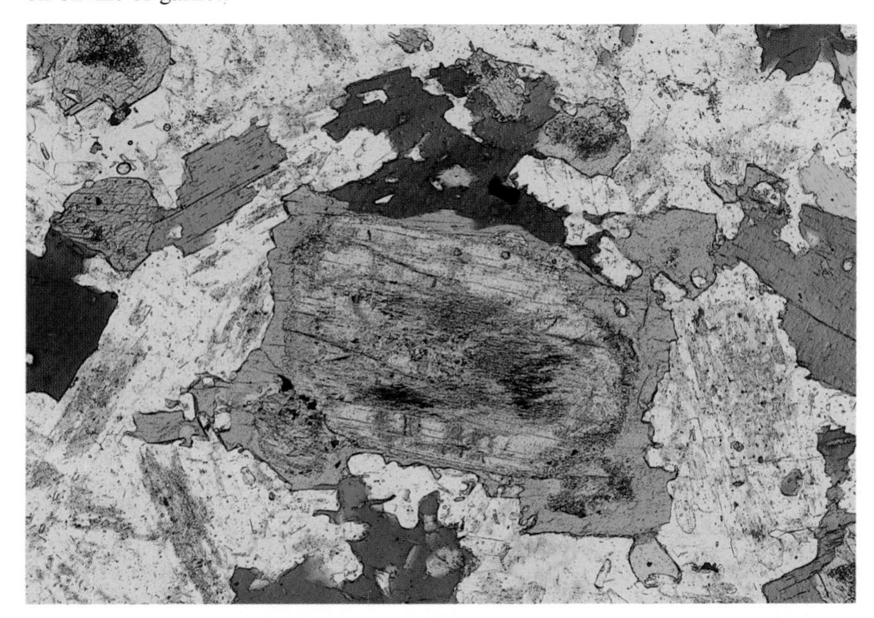

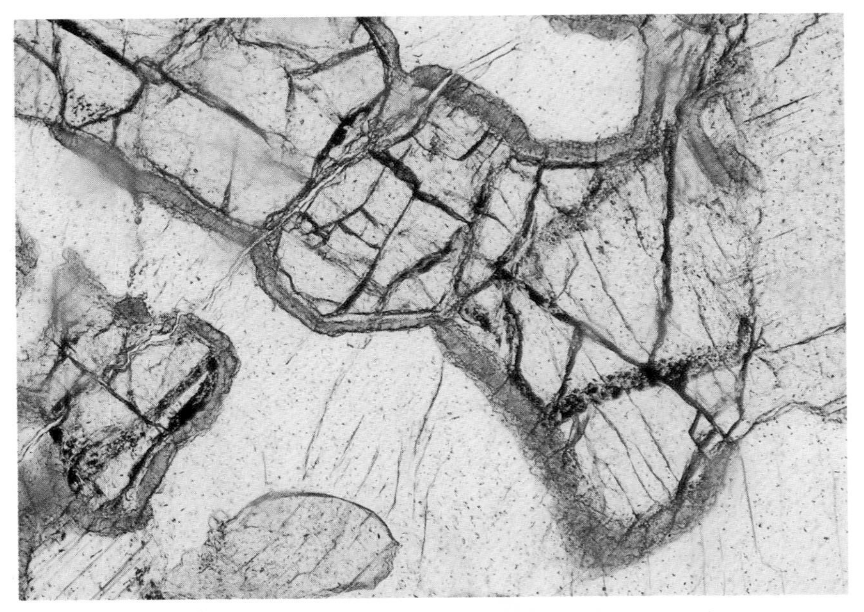

93 Corona texture

Between olivine and plagioclase crystals in this rock there is a 0.02–0.06 mm wide corona which consists of either one or two zones: (1) radially oriented, fibrous, brown hornblende; or (2) colourless pyroxene (see middle of photograph) surrounded by radially oriented, fibrous, brown hornblende. Analysis of the pyroxene suggests that it is a submicroscopic intergrowth of augite and orthopyroxene.

Olivine gabbro from Thessaloniki, North Greece; magnification × 100, PPL and XPL.

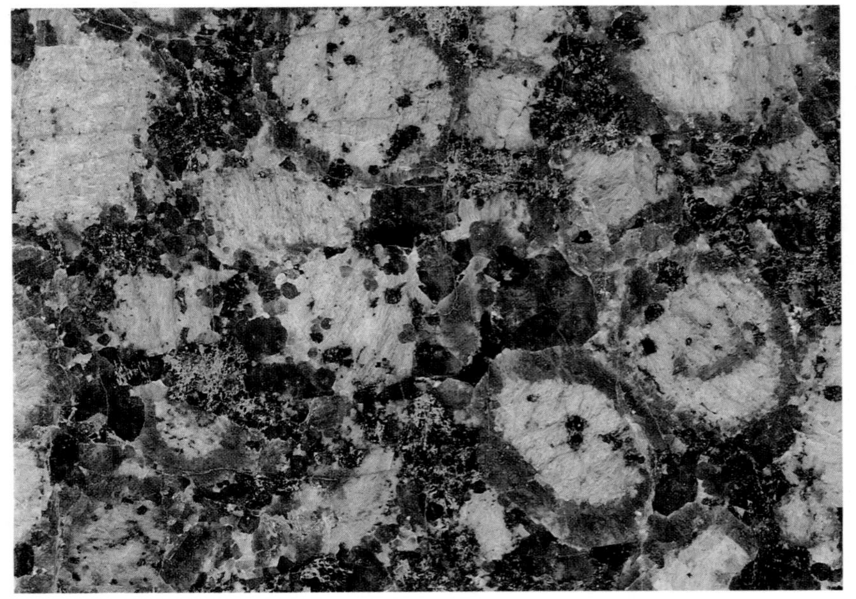

94 Rapikivi texture

The texture is of large, round potassic feldspars, some of which are mantled by sodic plagioclase rims, others have no plagioclase rims. In the first photograph, which is of a polished hand specimen, the plagioclase rims have a greenish colour contrasting with the pink potassic feldspar. The second photograph is of a thin section of the same rock.

Granite from Eastern Finland; magnification × 2 (first photo); × 3, XPL (second photo).

Rapakivi texture (continued)

Crystal zoning

One or more concentric bands in a single crystal are picked out by lines of inclusions (**95**) or by gradual or abrupt changes in solid-solution composition of the crystal. As regards the latter type of zoning, a large number of patterns are possible, the commoner ones being illustrated graphically and named below, using plagioclase as an example.

Normal versus reverse zoning
These terms specify the general trend of solid-solution composition from core to rim. 'Normal' indicates high-temperature component→low-temperature component (e.g. An-rich plagioclase→Ab-rich plagioclase, see Fig. C) and 'reverse' indicates the opposite.

Continuous[1] *versus discontinuous*[1] *zoning*
These terms indicate respectively a gradual or an abrupt change in composition. Figure C shows examples of *continuous normal zoning* and Fig. D an example of *discontinuous normal zoning*. Continuous and discontinuous zoning may alternate (Fig. E).

[1] *These terms are not the same as* continuous reaction *and* discontinuous reaction *of crystals with melt.*

95 Zonal arrangement of melt inclusions in plagioclase

Several stages in the growth of this plagioclase crystal can be picked out by the bands of minute melt inclusions. (See also **45**.)

Feldspar-phyric dolerite from Isle of Skye, Scotland; magnification × 9, PPL.

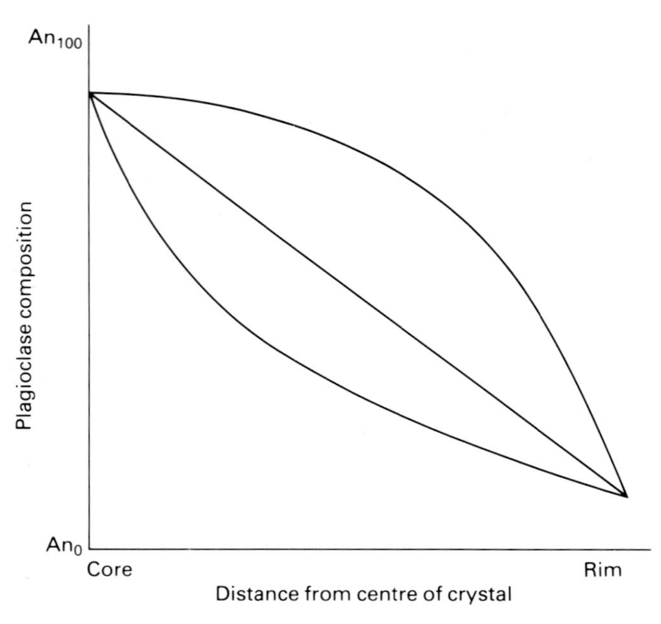

Fig. C Three examples of continuous normal zoning represented on a sketch graph

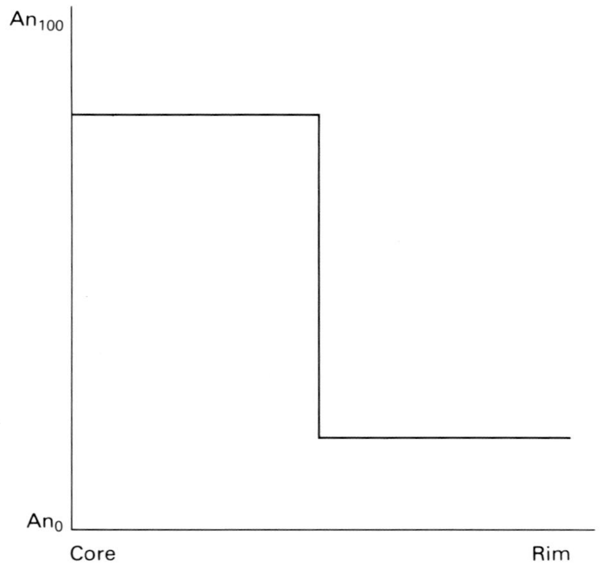

Fig. D Discontinuous normal zoning

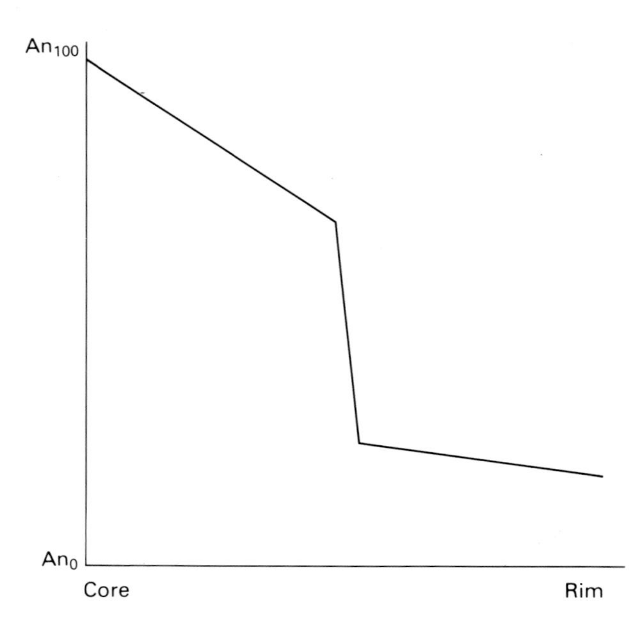

Fig. E Combined continuous and discontinuous normal zoning

96 Zoned plagioclase

The central plagioclase phenocryst in this photograph is discontinuously zoned, having a homogeneous core mantled by a more sodic rim; the rim has continuous normal zoning resulting in variation of the extinction position on rotation of the microscope stage. The crystal is thus partly discontinuously and partly continuously zoned.

Dolerite from Isle of Skye, Scotland; magnification × 43, XPL.

Multiple zoning

This term is used for crystals having repeated discontinuous zones. If the zones show a rhythmic repetition of width, the pattern is known as *oscillatory zoning*. The overall compositional trend of the multiple zoning may be *normal* or *reverse* or *even* (in which there is no general trend from core to rim). Individual zones may be of uniform or variable composition, such that the zoning pattern on a composition-distance graph is square wave, step-like, saw-tooth, curved saw-tooth, or some combination of these (see Figs. H–J). However, these are details which only *very* careful and lengthy optical examination or electron-probe microanalysis would reveal.

The reader should appreciate that the sketches in figs. C–J are all idealized and that in real crystals the oscillations will be less uniform; furthermore multiple or oscillatory zoning may only occupy part of a crystal, the remainder perhaps being homogeneous or continuously zoned.

Fig. F Multiple, even zoning

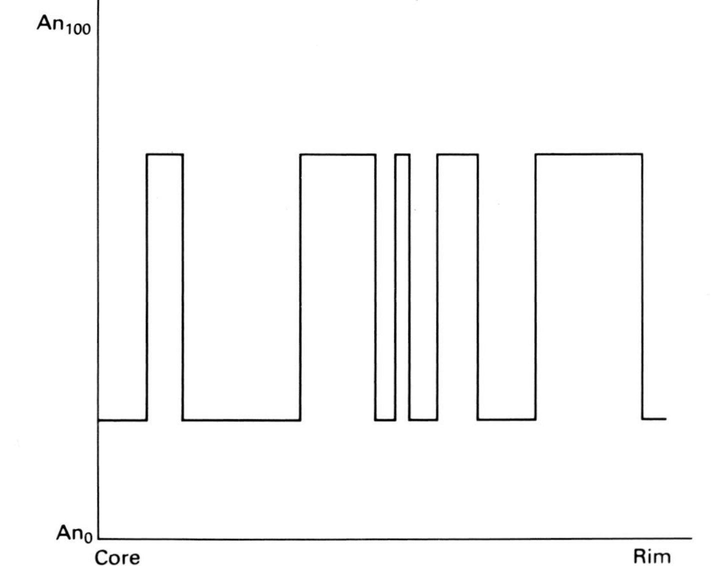

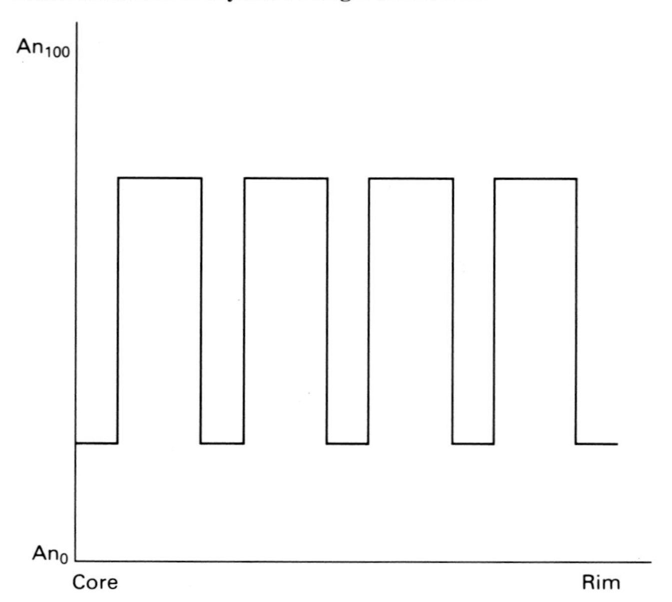

Fig. G Oscillatory, even zoning

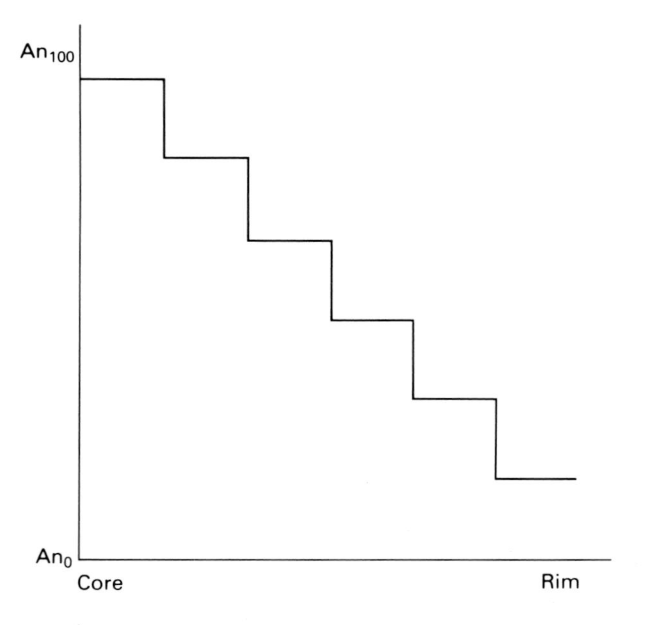

Fig. H Oscillatory, normal zoning: step-like

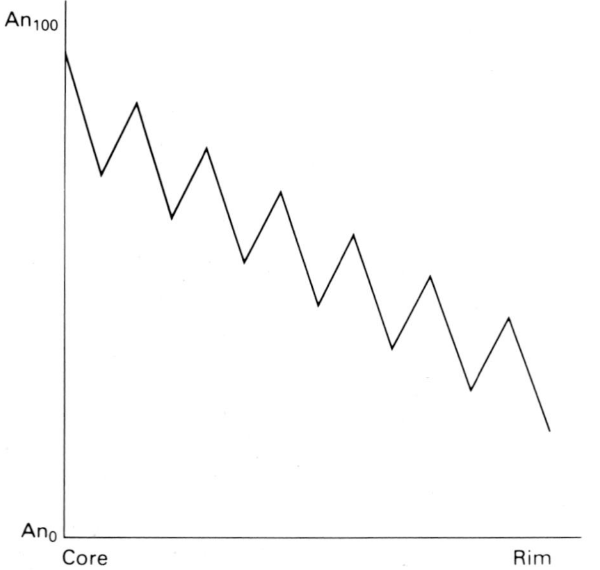

Fig. I Oscillatory, normal zoning: saw-tooth

Fig. J Oscillatory, normal zoning: curved saw-tooth

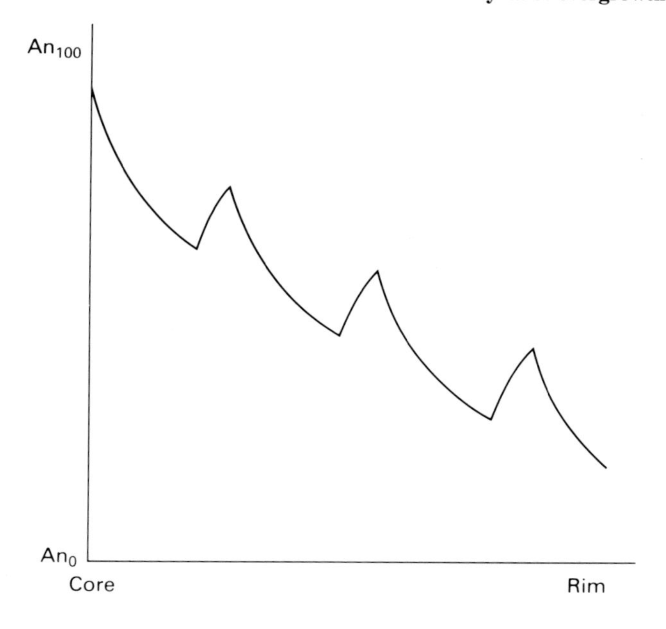

Convolute zoning

This is a variety of multiple zoning in which some of the zones are erratic and have non-uniform thickness (see **97**).

97 Zoned plagioclases

This photograph illustrates several styles of zoning in the two plagioclases comprising the glomerocryst. Combinations of discontinuous, oscillatory and convolute zoning are present, together with zoning picked out by a band of melt inclusions near the margins of both crystals.

Porphyritic andesite from Hakone volcano, Japan; magnification × 24, XPL.

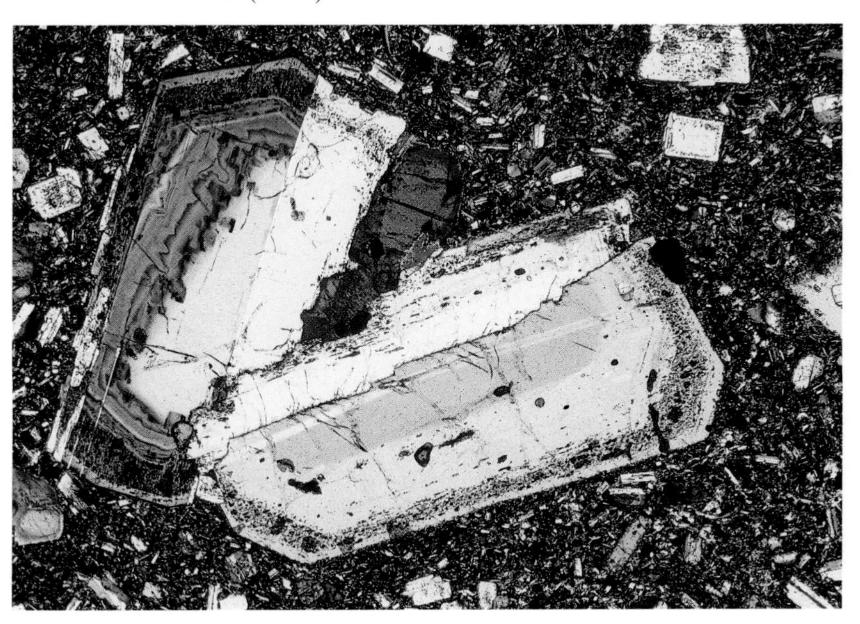

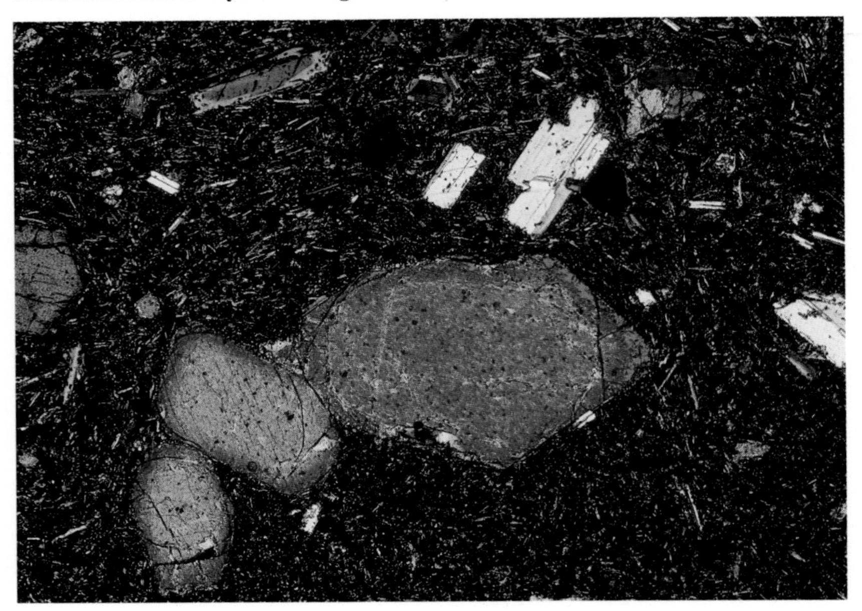

98 Zoned olivines

Zoning is not confined to feldspar crystals. Here, each of the three olivine phenocrysts in the cluster has a homogeneous core surrounded by a continuously normal-zoned mantle, as indicated by the variation in interference colours.

Ankaramite from Mauna Kea, Hawaii; magnification × 43, XPL.

Sector (or hourglass) zoning
As seen in thin section, this ideally takes the form of four triangular segments (sectors) with a common apex (Fig. K(b)). Opposite sectors are chemically identical, whereas adjacent ones differ in composition (though possibly only slightly) and hence in optical properties. Each sector may be homogeneous or show continuous or discontinuous or oscillatory, normal or reverse or even zoning. In three dimensions the sectors are pyramid shaped (Fig. K(a)), and, depending on the orientation of the crystal with respect to the plane of a thin section, a variety of patterns may be seen in thin section (Fig. K(b)–(f)). If the sector boundaries are curved, the pattern can resemble that of an hourglass (Fig. K(g)). Sector zoning is a common feature of pyroxenes in alkali-rich basic and ultrabasic rocks. It has also been seen in plagioclases in a few quickly cooled basalts.

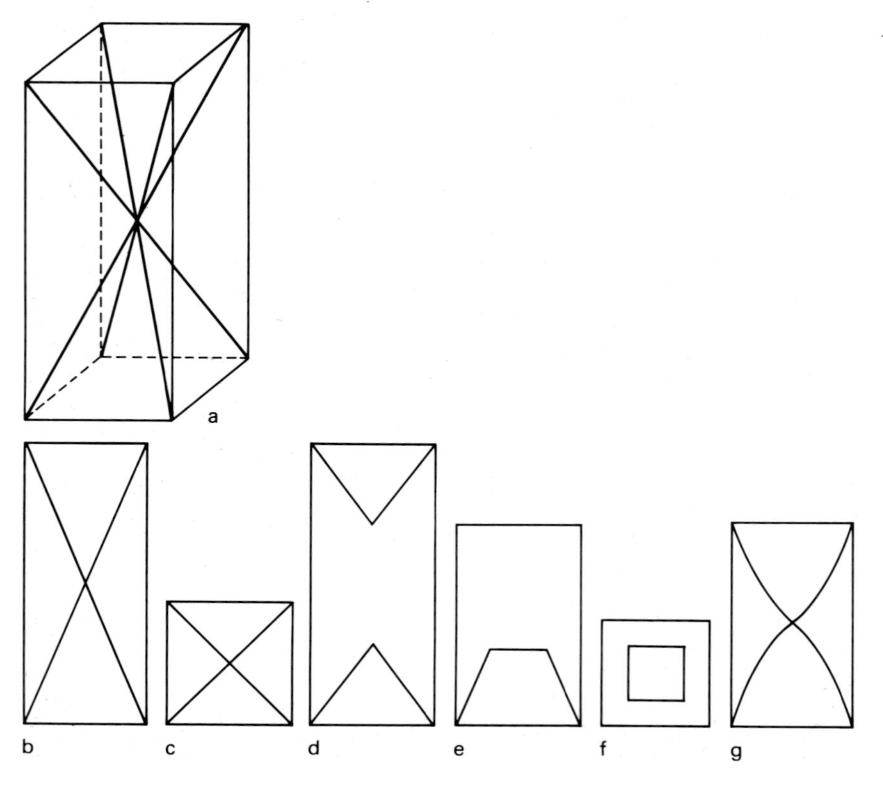

Fig. K Schematic representations of sector zoning

99 Sector-zoned augite

The picture shows a simple sector-zoned augite phenocryst containing elongate melt inclusions; the crystals partially enclosed by two of the sectors are olivines.

Essexite from Crawfordjohn, Scotland; magnification × 40, XPL.

100 Sector-zoned pyroxenes

Two sector-zoned titanaugite crystals are illustrated in these photographs; that on the left is complicated by forming at one end a graphic intergrowth with nepheline and leucite; the other crystal has an intriguing figure-8-shaped core, with a discontinuous, sector-zoned mantle.

Melanocratic nepheline microsyenite from Vogelsberg, West Germany; magnification × 7, PPL and XPL.

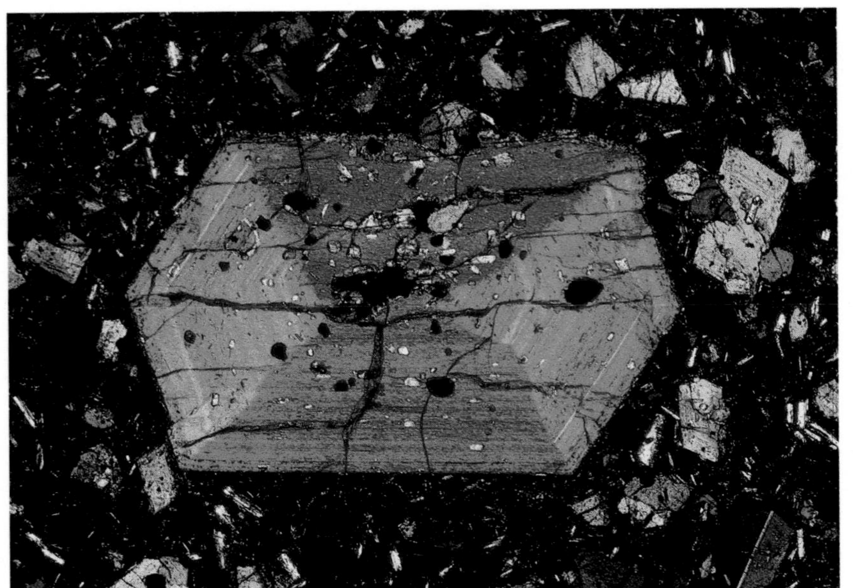

101 Oscillatory- and sector-zoned, inclusion-bearing pyroxene

The augite phenocryst occupying most of this photograph is sector-zoned and each sector displays oscillatory zoning. Inclusions of nepheline, augite and magnetite are arranged in trains parallel to the oscillatory zones.

Tephrite from Monte Vulturi, Malfi, Italy; magnification ×27, XPL.

102 Oscillatory- and sector-zoned pyroxene

Unlike the pyroxenes in **99 and 100**, this sector-zoned pyroxene has some sectors bounded by more than one face, e.g. the sector on the right is terminated by two faces, and that on the left by three faces. The crystal encloses plagioclase laths, an olivine (blue colour) and a pyroxene crystal (orange colour).

Essexite from Crawfordjohn, Scotland; magnification × 25, XPL.

Banded textures (banding)

Textures of this type involve two, or more, narrow (up to a few centimetres), sub-parallel bands in a rock which are distinguishable by differences in texture, and/or colour and/or mineral proportions. The term *layering* is also used by petrologists; while it includes banded texture, it is also used for larger scale stratification. An example of banded texture due to textural differences is illustrated in **5**, and **103** and **104** show examples resulting from extreme differences in mineral proportions.

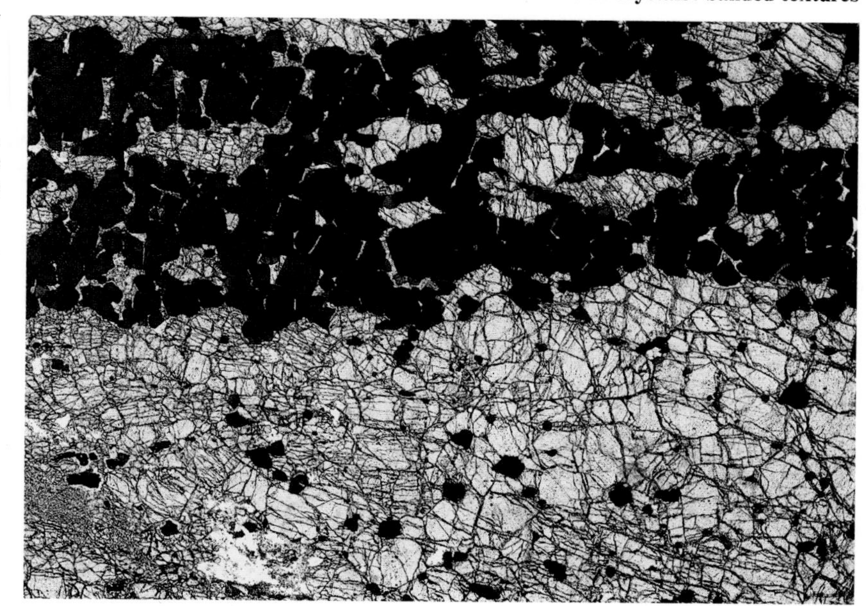

103 Olivine and chrome-spinel banding (or layering)

The photograph shows two bands, one rich in olivine, with scarce disseminated chrome-spinel crystals, and the other rich in equant chrome-spinel crystals with scarce interstitial olivine.

Banded dunite-chromitite from Skye, Scotland; magnification × 11, PPL.

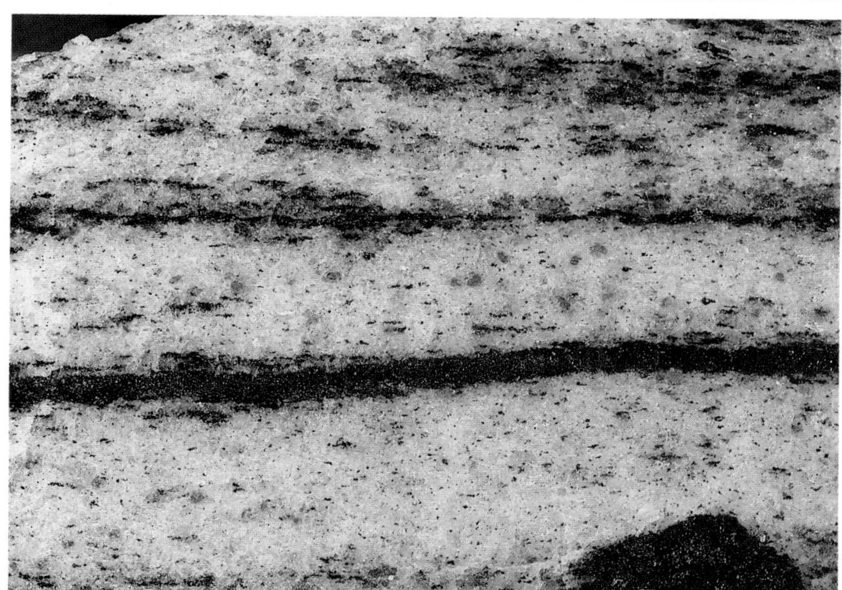

104 Anorthosite-chromitite banding (or layering)

This hand-specimen photograph shows alternating bands of anorthosite (white) and chromitite (black). The yellowish-brown crystals in the anorthosite are enstatite and the black particles are single crystals and glomerocrysts of chromite.

Banded anorthosite-chromitite from Critical Zone of the Bushveld intrusion, South Africa; magnification × 2.

Comb layering, orbicular texture, ocellar texture and eutaxitic texture

Comb layering (see p. 44, **70** and **71**) and *orbicular texture* (**105**) are particularly exotic kinds of banding. In the latter, 'orbs' consist of concentric shells of rhythmically alternating mineral constitution. Within the shells the texture may either be granular or elongate crystals may be radially arranged. 'Orbs' may reach a few tens of centimetres in diameter. A further variety of banded texture, *eutaxitic*, occurs in some tuffs and ignimbrites and consists of a regular alignment of flattened glassy fragments (**8b**).

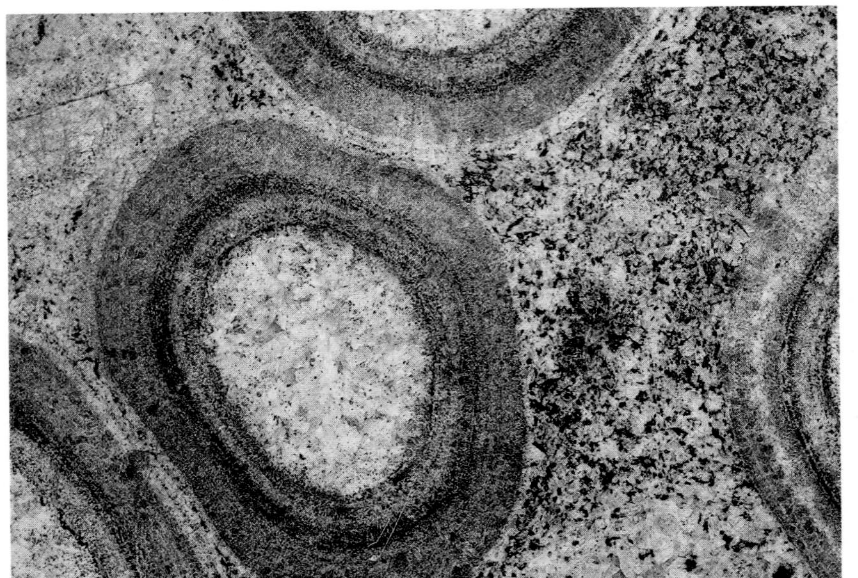

105 Orbicular monzodiorite

The first photograph shows the texture in a hand specimen. The arrangement of the concentric darker bands about the lighter coloured, homogeneous nuclei is well displayed. The second photograph shows the core and a few inner bands of one orbicule in thin section. The bands can be seen to differ from one another in their contents of biotite and alkali feldspar, and in their grain size.

Monzodiorite from the Island of Suuri Lintusaari, Ruokolahti, S.E. Finland; magnification × 1 (first photo) × 3, and XPL (second photo).

Cavity textures

These are a collection of textures which feature either holes in the rock or likely former holes which are now partly or completely filled with crystals.

Vesicular texture

Round, ovoid, or elongate irregular holes (vesicles) formed by expansion of gas, in a magma.

Amygdaloidal texture

Former vesicles are here occupied, or partially occupied, by late-stage magmatic and/or post-magmatic minerals, such as carbonate, zeolites, quartz, chalcedony, analcite, chlorite, and/or, rarely, glass or fine groundmass. The filled holes are known as amygdales or amygdules.

Ocellar texture

Certain spherical or ellipsoidal leucocratic patches enclosed in a more mafic host are known as ocelli (singular ocellus). Unlike amygdales, the minerals filling an ocellus can normally all be found in the host rock; they may include any of: nepheline, analcite, zeolites, calcite, leucite, potassium feldspar, sodium feldspar, quartz, chlorite, biotite, hornblende and pyroxene, or even glass, and the minerals are commonly distributed in a zonal arrangement (**109a**). Often, platy and acicular crystals in the host bordering an ocellus are tangentially arranged (as in **109b**) but sometimes project into the ocellus. Ocelli are normally less than 5 mm in diameter but may reach 2 cm. Their origin has been ascribed on the one hand to separation of droplets of immiscible liquid from magma, and on the other hand to seepage of residual liquid or fluid into vesicles.

Miarolitic texture

These are irregularly shaped cavities (druses) in plutonic and hypabyssal rocks into which euhedral crystals of the rock project.

Lithophysa (or stone-ball)

This is the term given to a sphere consisting of concentric shells with hollow interspaces.

106 Vesicular feldspar-phyric basalt

Large subspherical gas cavities are randomly distributed in this volcanic rock. Note the two vesicles at the top left which have coalesced.

Basalt from Mount Fuji, Japan; magnification × 7, PPL and XPL.

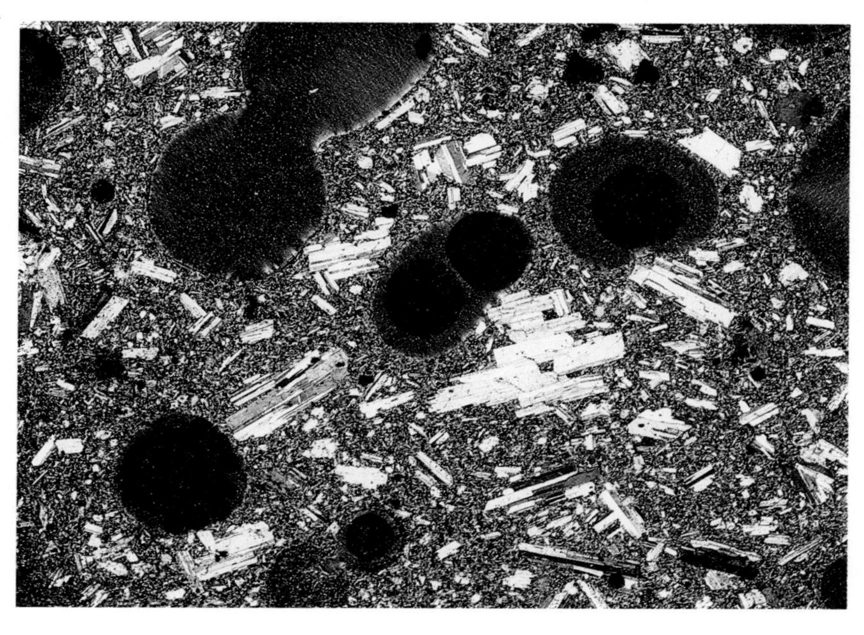

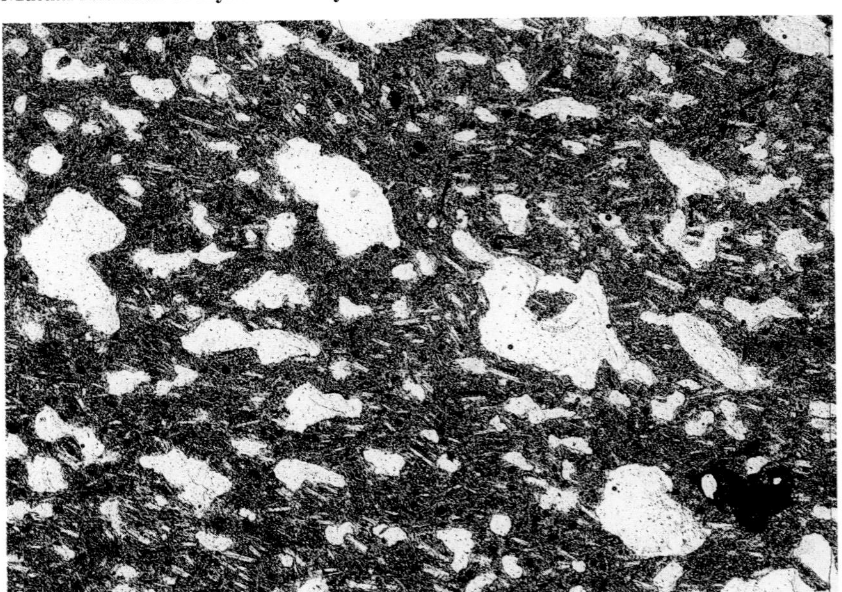

107 Vesicular trachyte

Irregularly shaped, elongate vesicles are streaked out through this trachyte; the columnar feldspars show a weak alignment in the same direction.

Trachyte from the Auvergne, France; magnification × 12, PPL.

108 Amygdaloidal basalt

The original vesicles in this volcanic rock are now filled with an aggregate of small calcite crystals; calcite is also present as pseudomorphs after olivine in the groundmass. Pyroxene and glass in the rock are altered to clay minerals.

Basalt from Matlock, Derbyshire, England; magnification × 11, PPL and XPL.

124 *shows another amygdaloidal rock.*

109 Ocellar texture

The upper photograph shows three ocelli in an olivine dolerite sill. Each ocellus is outlined by a more or less complete veneer of tiny magnetite crystals. At the base of the two largest ocelli the groundmass outside the ocelli extends across the magnetite veneer, except that olivine is absent inside the ocelli. The remainder of each ocellus comprises clear zeolite, turbid, very fine-grained zeolite and scarce magnetite. The left-hand ocellus also contains three elongate pyroxenes on the left side.

Non-porphyritic facies of an olivine dolerite sill, Igdlorssuit, Ubekendt Ejland, West Greenland; magnification × 12, PPL.

The second picture shows two ocelli, occupied by calcite, alkali feldspar, chlorite and fine-grained patches of clay (possibly altered glass). Laths of biotite are arranged tangentially about each ocellus.

Minette from Westmorland, England; magnification × 16, PPL.

109a Miarolitic (or drusy) cavity in granite

The third photograph shows miarolitic (or drusy) cavity in granite. The slightly angular cavity shown in this hand specimen is occupied by crystals of alkali feldspar, quartz and biotite, some up to seven times larger than crystals of the same minerals in the rest of the granite.

Granite from Beinn an Dhubaich, Skye, Scotland; magnification × 1.5.

Part 2

Varieties of igneous rocks

Introduction

In this Part are defined and illustrated many of the more common igneous rock types. For most types plane-polarized light and crossed-polarized light views are shown. In a few cases more than one example of the rock type is illustrated and in some we have used more than one magnification to show a particular feature of a rock. In addition, reference is made to Part 1 on textures where other examples of a specific rock type are illustrated. Thus, although gabbro may be represented by only two photographs in Part 2, we have noted where other photographs of gabbros appear in Part 1.

No two igneous rocks are identical in every respect but many are sufficiently alike that they can be illustrated by a few typical specimens. Thus, an olivine gabbro from one locality may be very similar to a large number of olivine gabbros from different parts of the world. We have therefore selected thin sections which are fairly typical of the rock type being illustrated.

The choice of which rocks to include has not been easy and undoubtedly we have omitted somebody's favourite. In Johannsen's *Descriptive Petrography of the Igneous Rocks* more than 540 different names for igneous rocks are listed in the index, not counting those names which have a prefix indicating the presence of a particular mineral or texture, thus we have counted *diorite* as one name rather than the eighteen varieties of diorite listed by Johannsen. Holmes listed about 340 different igneous rocks in his *Nomenclature of Petrology* but probably less than 150 of these are now in common use. We have selected about sixty of these names as rocks which the student may expect to see in an undergraduate course in geology. Certain rock types cannot be distinguished by a cursory examination of a thin section, much less from one or two photographs. Thus, for example, because *mugearites* and *hawaiites* cannot readily be distinguished from *alkali basalts* without a determination of the plagioclase composition, photographs of these rocks in thin section have not been included.

For each rock illustrated we have given what we consider to be the definition of

Fig. L Nomenclature of the commoner igneous rocks based on their silica and alkali contents.

Names of fine grained rocks are shown in small letters and those of coarse-grained rocks in capital letters (modified from Cox, Bell and Pankhurst, 1979)

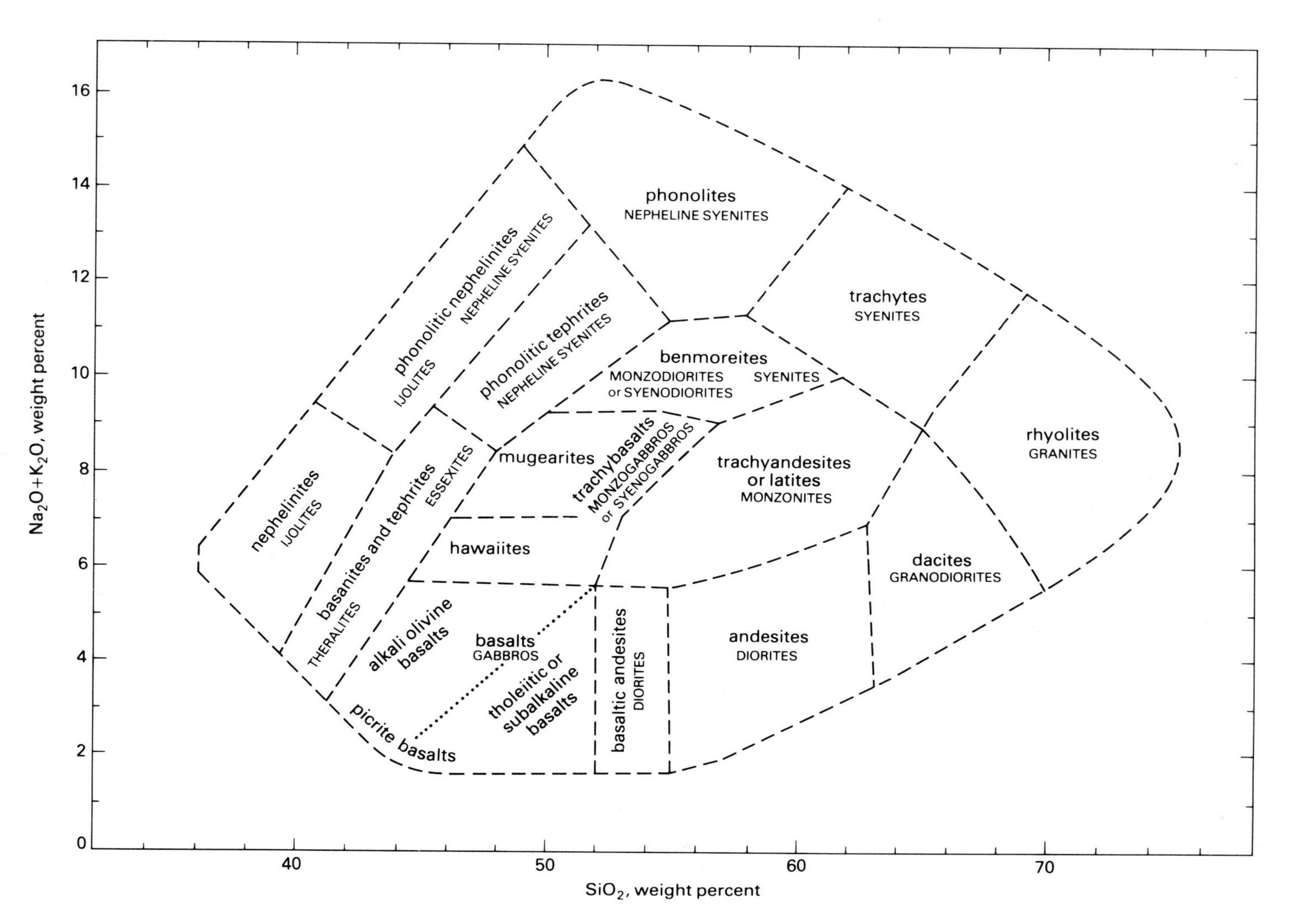

the name and this is followed by a brief description of what is visible in the field of view of the photomicrographs. In addition to defining the rock illustrated, we have also defined, though not necessarily illustrated, the names of others which are subtypes and whose names are still in use (e.g. *granophyre* as a variety of *microgranite*). Agreement among petrologists on the characteristics of individual rock types is improving but will always be open to some differences of opinion. The names used and defined here are as near to consensus opinion as we can sense it, using the text of Holmes (1920), Johannsen (1931), Hatch *et al.* (1972), and Nockolds *et al.* (1978), the paper of Wilkinson (1968), and our own experience. In most of the definitions we have refrained from stating ranges for the amounts of the essential minerals, since agreement amongst petrologists on this is generally poor. On the other hand, the photographs give the reader an indication of whether a particular mineral is abundant or scarce in the rock.

The photograph descriptions are deliberately short because they are only of those features which can be seen in the photomicrographs. The complete petrographic description of a rock requires a careful examination of the whole slide at different magnifications and the student is likely to see much more than can be illustrated in one view at one magnification.

We have not set out a system for the classification of igneous rocks because this is beyond the scope of this book. The sequence in which the rocks are arranged is broadly ultrabasic and basic rocks first, followed by intermediate and then acid rocks, leaving the alkali-rich rocks to the end. Among the alkali-rich rocks are included a number of rare rock types, simply because they are rare and because the photographs are visually attractive. In any treatment of petrography those rocks traditionally grouped together under the name *lamprophyres* pose a problem because of their diverse characters. Although we have defined some of them, we have illustrated only three – *minette*, *alnöite* and *fourchite*.

While we have avoided a formal classification scheme of rocks, it is nonetheless helpful to have in the mind's eye a series of pigeon holes in which to locate rock names with respect to one another; otherwise the brain tends to succumb to the weight of names and refuses to accept more than a few of them. Figure L (p. 77), modified from Cox, Bell and Pankhurst (1979), is a chemical diagram on which many rock compositions may be plotted. The outermost line encloses most known volcanic rocks and the bounded area has been subdivided and the names of fine-grained and coarse-grained varieties of rocks indicated. The exact positions of the dividing lines and the names in each area are open to debate but, in general, most petrologists would accept this classification. While a great many of the rock types illustrated here are shown on Fig. L, a small proportion are not – e.g. the names on the figure apply to the chemical condition in which Na is less than K. Other names are used for rocks with the much less common condition of K greater than Na (e.g. *leucitite* instead of *nephelinite*).

In the photograph descriptions a number of terms are used which are worthy of definition here:

Essential minerals: those which are necessary to the naming of the rock. They need not be major constituents, e.g. a *crinanite* contains only a small percentage of essential analcite.

Accessory minerals: those which are present in such small amounts in a rock that they are disregarded in its definition, e.g. a small percentage of quartz in a gabbro. However, it may be useful in the name to note the presence of a particular accessory mineral in a rock and this can be done by adding the mineral name as a prefix, e.g. *quartz gabbro*.

Melanocratic, mesocratic and leucocratic (synonymous with *dark-*, *medium-*, and *light-coloured*): terms to indicate the colour index of a rock and hence the relative proportions of dark- to light-coloured minerals. The boundaries are at 66% and 33% dark minerals respectively. *Mafic* and *felsic* may be applied to rocks which are composed predominantly of mafic minerals (olivine, pyroxenes, amphiboles, biotite, opaque minerals) or of felsic minerals (quartz, feldspar and feldspathoid), respectively. They are thus less precise than the colour index terms. The term *ultramafic* is used for rocks with trivial amounts of, or no, felsic minerals. The rarely-used colour index term *hypermelanic* (90–100% dark minerals) is more or less equivalent to ultramafic.

Ultrabasic, basic, intermediate and acid: chemical terms to designate rocks with less than 45%, 45–52%, 52–66% and more than 66% by weight of SiO_2 respectively. Since a large SiO_2 content is reflected in a large amount of light-coloured minerals, these terms correspond approximately to the colour index ones.

Micro as a prefix: most igneous rocks have fine-, medium-, and coarse-grained varieties. The fine- and coarse-grained varieties always have different names (e.g. basalt and gabbro). Medium-grained varieties may also have a distinct name (e.g. dolerite), or more often these days, the name for the coarse-grained rock is used and prefixed by *micro* (e.g. microgranite, microsyenite or even microgabbro).

110

Dunite

This is the name used for an ultramafic rock which consists almost entirely of olivine, often accompanied by accessory chrome spinel.

The granular-textured sample we have illustrated consists of only two minerals, olivine and a chromium-rich spinel. The spinel appears opaque in the PPL view but, with a more intense light than can be used for photography, it can be seen to have a deep brown colour. A banded structure is visible in the large crystal showing a blue interference colour to the right and slightly up from the centre of the photograph, and in two of the crystals showing brown interference colours to the right of the crystal showing blue. Above the blue crystal a crystal shows irregular extinction. These features indicate that the olivines are strained.

Dunite from Mount Dun, New Zealand; magnification × 16, PPL and XPL. Another dunite is illustrated in ***103***.

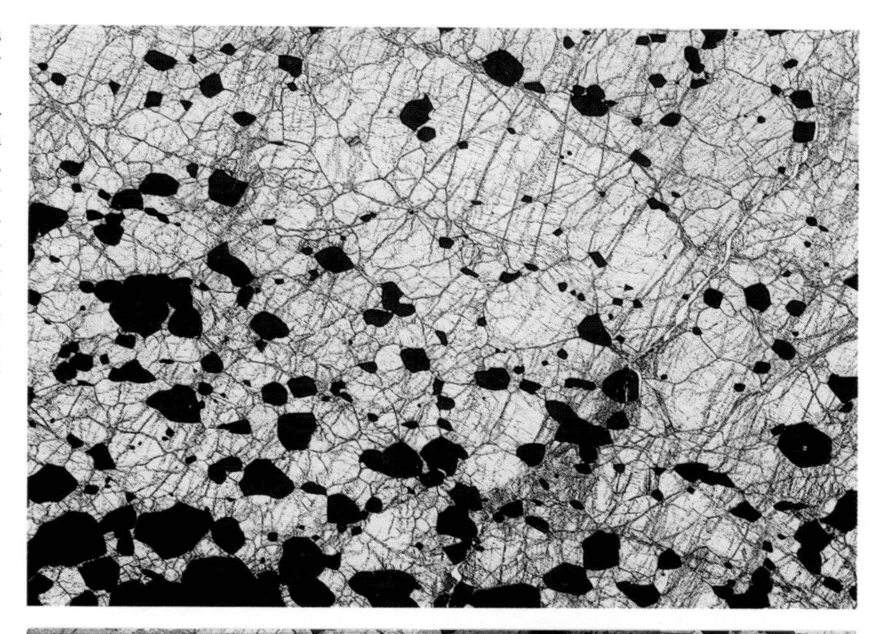

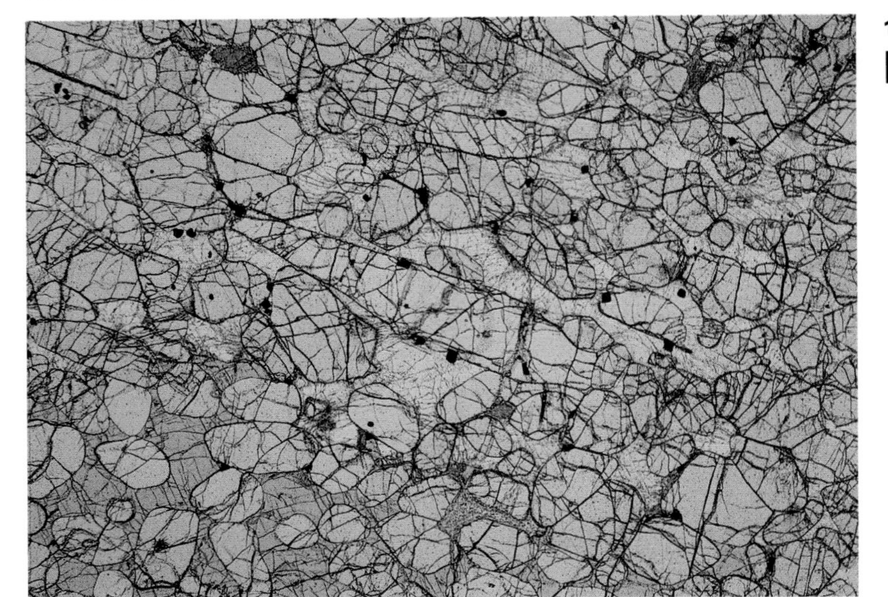

111 Peridotite

This term is used for coarse-grained olivine-rich rocks in which olivine is the dominant mineral but is less than 90% of the rock. Textbooks usually state that the accessory minerals are all ferromagnesians and that olivine-rich rocks containing plagioclase and pyroxene should be called *picrites* (or *troctolite*, if olivine and plagioclase only), (**41**, **51**). However, picrite is not much used now, and modern usage allows for plagioclase to be present in peridotite, as can be indicated by the terms *plagioclase* or *feldspathic peridotite* as in **51**. Peridotites containing both orthopyroxene and clinopyroxene (**113**) are often called *lherzolites*. If clinopyroxene is present and orthopyroxene in a minor amount or absent, *wehrlite* is used, and *harzburgite* for the converse.

We have chosen to illustrate this rock by two different samples.

The upper and middle photographs show a poikilitic textured peridotite in which, in the lower left part of the field, a number of round crystals of olivine are embedded in two clinopyroxene crystals, and elsewhere the olivines are enclosed by plagioclase crystals. In the centre of the field, one elongated olivine crystal is surrounded by plagioclase feldspar. The small opaque crystals are chromite. The differences in colour and relief between the plagioclase and the pyroxene are also obvious in the PPL view: stray polarization produces the pale greenish and pink colours in this view.

The lower photograph shows an XPL view of a peridotite in which numerous olivine crystals are poikilitically enclosed in a basic plagioclase feldspar. Only a small proportion of pyroxene is present in this rock.

First and second photographs: Peridotite from Rhum, Scotland; magnification × 12, PPL and XPL.
Third photograph: Peridotite from the Shiant Isles, Scotland; magnification × 15, XPL.
*Other peridotites are illustrated in **18** and **48**.*

112
Kimberlite

Kimberlite is a porphyritic potassium-, water-, and carbon dioxide-rich peridotite which forms dykes, sills and pipes. It consists of phenocrysts of olivine, phlogopite, ilmenite and pyrope garnet in a groundmass which commonly contains olivine, phlogopite, serpentine, calcite, chlorite, magnetite, apatite and perovskite.

All of the phenocrysts in the photographs are olivine, as are many of the smaller crystals; some of the small crystals are pyroxene. The olivine crystals have round outlines and are surrounded by rims of microcrystalline serpentine. In the groundmass are patches of calcite and a high density of small crystals which appear black in the photographs: these are oxide minerals, including perovskite ($CaTiO_3$) which is a common constituent of kimberlites. The large olivine at the top left contains neoblasts, i.e. new, smaller crystals which are believed to have grown from highly strained parts of the big crystal.

Kimberlite from Kimberley, South Africa; magnification ×7, PPL and XPL.

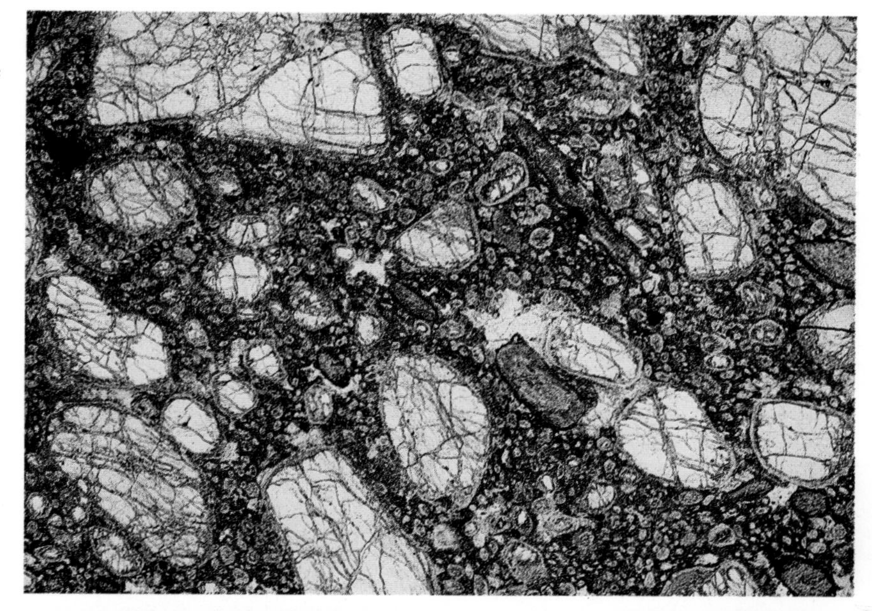

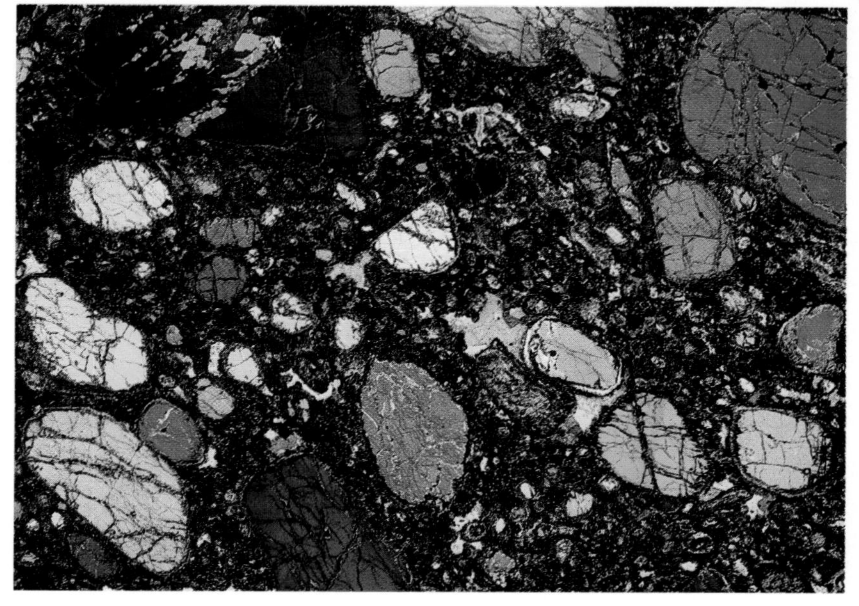

113
Garnet peridotite

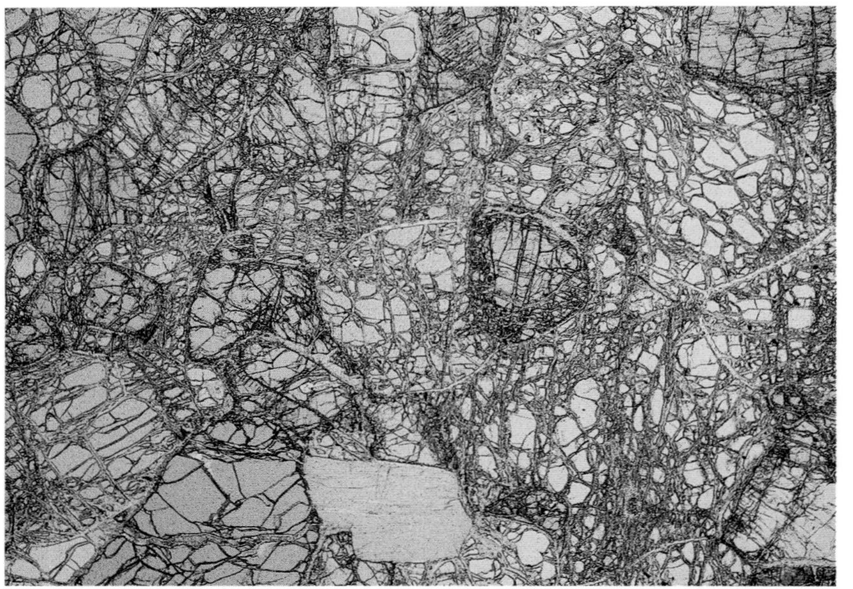

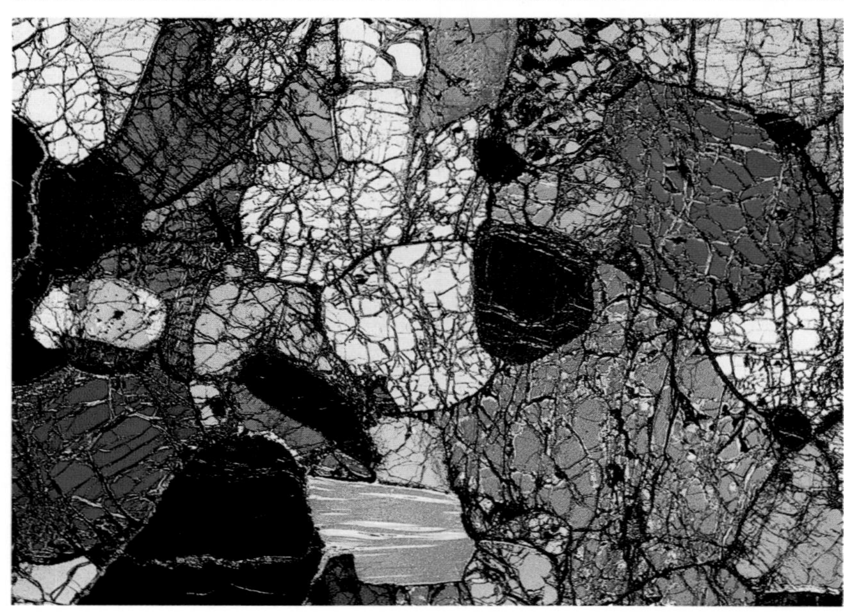

A plutonic rock consisting of more than 40% olivine with two kinds of pyroxenes and a small amount of garnet.

The granular-textured specimen illustrated has only one garnet fairly clearly shown in the field of view; it is at the left of the lower edge. About the centre of the lower edge is a crystal of strained phlogopitic mica. The rest of the field is occupied by olivine (grey), orthopyroxene (pinkish-brown), and chrome diopside (pale green e.g. middle of top edge), the olivines and chrome diopside showing moderate interference colours, orthopyroxene showing low grey colours in the XPL view.

The veins which penetrate most of the crystals are of serpentine but around the garnet crystals are thin veins which are occupied by a pale brown mica: in the XPL view the interference colours of the micas are moderate second order.

Garnet peridotite from Kimberley, South Africa; magnification × 7, PPL and XPL.

114

Pyroxenite

An ultramafic rock consisting mainly of pyroxene(s); possible accessory minerals include olivine, spinel, garnet, hornblende, biotite, feldspar, and nepheline. A rock consisting of both orthopyroxene and clinopyroxene is known as a *websterite*. If aegerine-augite or nepheline is present the term *alkali pyroxenite* is used.

The sample illustrated here is, strictly speaking, a websterite and shows round orthopyroxene crystals, recognized by their low first-order colours and a lamellar structure, poikilitically enclosed by a large zoned clinopyroxene crystal showing blue and red interference colours. A few areas of low relief seen in the view in PPL are basic plagioclase. At the bottom left is a hole in the slide. A slight difference in colours of the orthopyroxene crystals can be seen in the PPL view and this is chiefly due to stray polarization in the photographic equipment.

Websterite from the Stillwater complex, Montana, USA; magnification × 7, PPL and XPL.

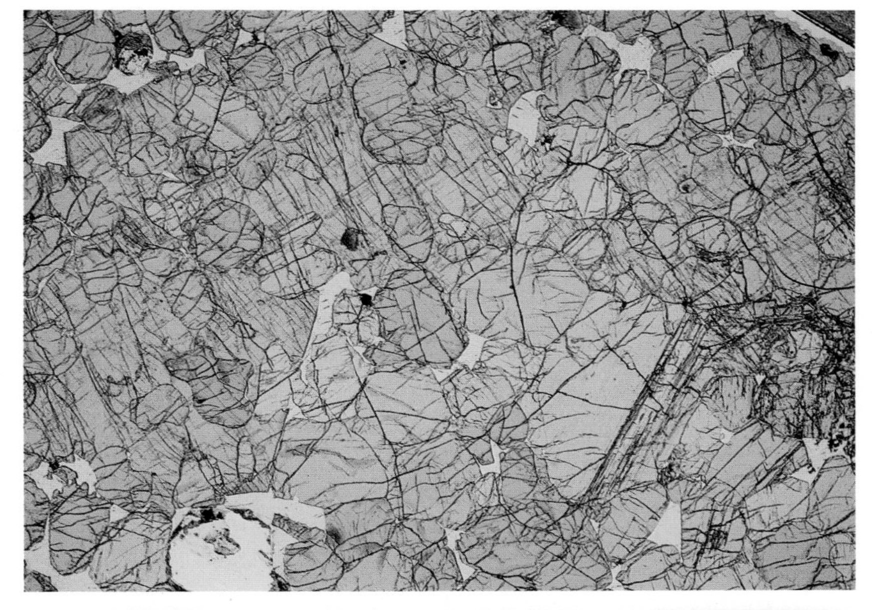

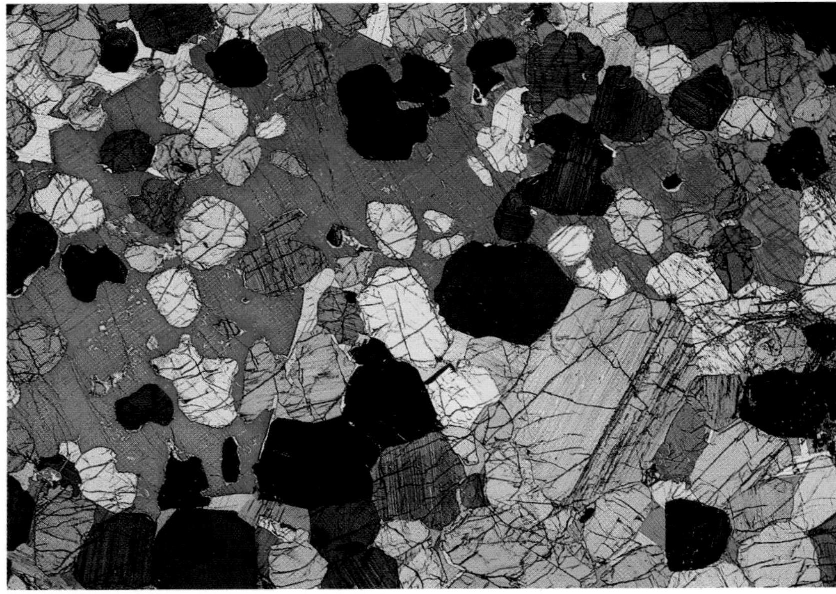

115
Komatiite

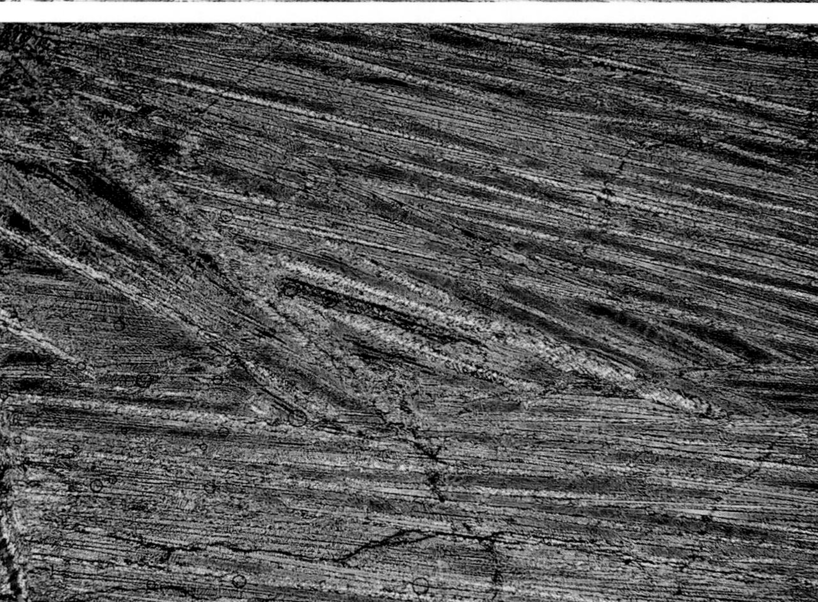

This has been defined as an ultramafic volcanic rock with more than 18% MgO. While the rock lacks plagioclase, it is rich in magnesian olivine and clinopyroxene crystals set in devitrified glass, though the olivine is usually all but completely serpentinized. The term 'spinifex textures' has been used to describe parallel and radiate arrangements of elongate olivines and pyroxenes common in komatiites. These textures imply rapid crystallization from an ultramafic liquid.

The photomicrographs were made from a large thin section (15 cm by 6 cm) in which pseudomorphs after olivine crystals are up to 10 cm in lengths. In the PPL view the long colourless shapes were originally olivines but are now completely replaced by serpentine. In thin section the olivine crystals appear to be acicular in habit but in fact are cross sections of thinly tabular crystals stacked parallel to one another. The brown material (PPL view) was mainly clinopyroxene and glass, though the clinopyroxene has been replaced by chlorite and tremolite or talc, and the glass replaced by chlorite.

Komatiite from Munro Township, Ontario; magnification ×3, PPL and XPL.

116
Meymechite

This is the name given by the Russians to a rock discovered in Siberia which is a porphyritic ultramafic extrusive rock consisting mainly of olivine phenocrysts in a groundmass of clinopyroxene, mica and chlorite.

The illustrated sample comes from the type locality and shows parts of two large phenocrysts of partially serpentinized olivine in a groundmass consisting mainly of brownish pyroxene and some iron ore. The fox-brown mineral, of which there are only small fragments, is mica; some chlorite is present. The groundmass is highly serpentinized.

Meymechite from the Meymecha river, Northern Siberia; magnification × 10, PPL and XPL.

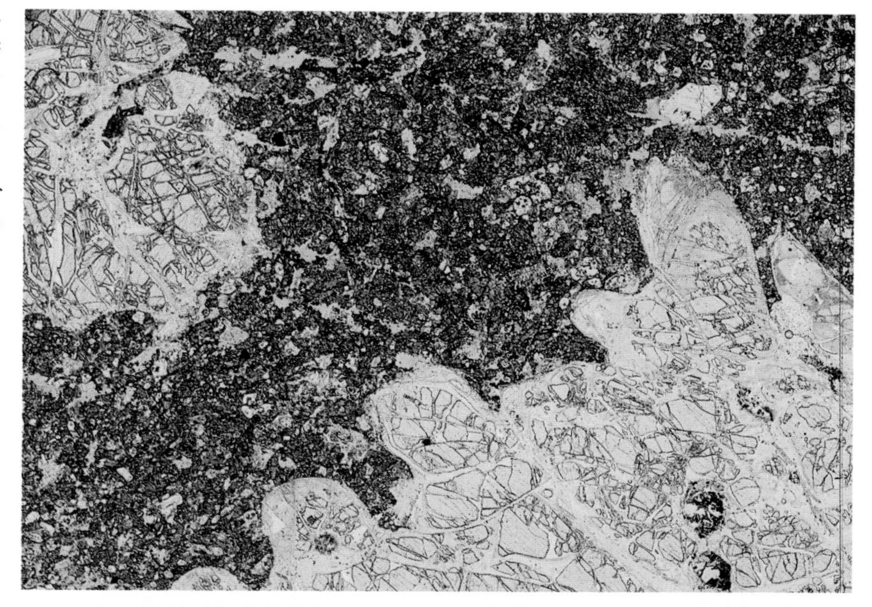

117
Hornblendite

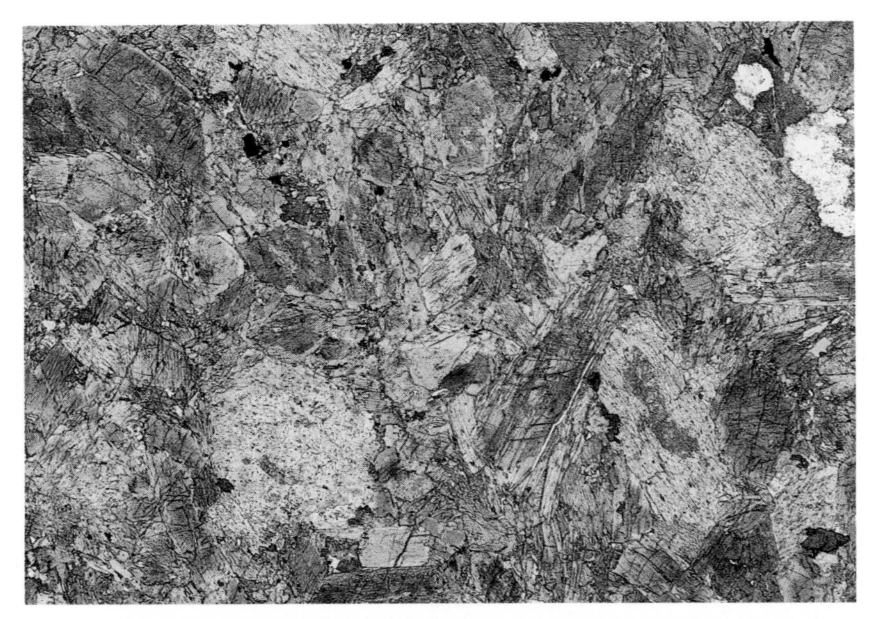

An ultramafic igneous rock consisting mainly of hornblende. The name *amphibolite* is reserved for a metamorphic rock consisting essentially of an amphibole and plagioclase.

The first and second photographs are of a hornblendite in which almost the whole of the field of view is occupied by amphibole crystals. Many of the crystals are zoned, a few are twinned and a few are cut in the correct orientation to show the two cleavages intersecting at 120°. Notice the lack of preferred orientation of the crystals and their interlocking relations.

Hornblendite from Donegal, Ireland; magnification × 12, PPL and XPL.
*Another hornblendite is illustrated in **39**.*

Basalts

In the most generalized definition these are fine-grained mafic rocks with essential augite, labradorite-sodic bytownite and opaque minerals (titanomagnetite ± ilmenite). They may be subdivided, if so wished, into *tholeiitic basalts* (*tholeiites* or *subalkaline basalts*) and *alkali olivine basalts* (fig. L) on the basis of the presence or absence of accessory olivine, quartz and low-Ca pyroxenes (pigeonite or orthopyroxene). Phenocrysts or microphenocrysts of all the essential and accessory minerals (except quartz) may be present.
Tholeiitic basalts (**118**, **46**, **58**, **62**) contain both augite and low-Ca pyroxene (pigeonite, hypersthene or both). Olivine is either absent or present only in small amounts (less than 5% by volume) as phenocrysts only, never in the groundmass. The groundmass commonly contains varying amounts of interstitial brown glass, or devitrified glass (intersertal texture); in more slowly cooled rocks the place of the glass is taken by granophyric intergrowths of quartz and alkali feldspar. The remainder of the groundmass usually has an intergranular or subophitic texture.
Alkali olivine basalts (**119** and **11**) contain no low-Ca pyroxene but plentiful olivine, both as phenocrysts (if present) and in the groundmass. The augite is often somewhat purplish-grey in colour due to high Ti content. Less than 10% of the feldspar is of alkali type. The groundmass texture is usually intergranular or subophitic and glass is very rare, though accessory interstitial nepheline or analcite may be present. If alkali feldspar is present, it is in the interstices and as rims on plagioclase.

The terms *olivine tholeiite* and *olivine basalt* (**22**, **23**, **44**, **56**, **57**) have been used for rocks which have certain characteristics of both tholeiites and alkali olivine basalts. They lack low-Ca pyroxene, olivine is essential, exceeds 5% and may be present as both phenocrysts and in the groundmass. The augite is not Ti-rich and it is not as Ca-rich as that in alkali olivine basalts. Interstitial glass may be present: nepheline and analcite are absent.

Chemical data greatly assist in making these distinctions: normative hypersthene is the hallmark of a tholeiite (true tholeiite and olivine tholeiite) and absence of normative hypersthene characterizes alkali olivine basalt; normative olivine and hypersthene characterize olivine tholeiite. Without such data the petrographer must rely on the mineralogical characteristics mentioned above, which of course may not be distinguishable if the rock is very fine grained.

Particularly olivine-rich varieties of both alkali olivine basalt and olivine tholeiite exist (up to 50% olivine) and these may be referred to as *alkalic picrites* and *tholeiitic picrites* (or *tholeiitic picrite basalts*) or generally *picritic basalts* (**122**, **26**, **27**, **31**). Pyroxene-rich basalt is called *ankaramite* (**98**, **123**).

The term *trachybasalt*[1] is sometimes used for rocks slightly richer in alkalis and silica than alkali olivine basalt and hence having a more sodic plagioclase and more alkali feldspar (10–40% of total feldspar) than alkali olivine basalt. Strictly, the term should be used for those rocks which on an alkali-silica plot (fig. L) lie between alkali olivine basalt and trachyte, namely *hawaiite* (andesine, anorthoclase, olivine, augite and biotite, see **47**), *mugearite* (same but oligoclase for andesine) and *benmoreite* (same but anorthoclase for oligoclase) and hence show features gradational between trachyte and basalt. *Syenogabbro* is the equivalent coarse-grained name.

A very uncommon group of basalts are both alkali rich and have K greater than Na, in contrast to common basalts. These contain essential K-feldspar in the groundmass in addition to augite, plagioclase (labradorite) and opaques. Olivine and biotite are common accessories. The terms *absarokite* and *shoshonite* are used for these, the former being more mafic than the latter.

Lunar basalts, two of which are illustrated here (**120**, **121**), are classified differently but being poor in sodium and potassium are more akin to terrestrial tholeiites than to alkali olivine basalts.

---

[1] *Consensus on the meaning of this term is poor – it was originally used for what is now termed a basanite (***157***) and some petrologists have used it for basalts in which the K content exceeds that of Na.*

118
Basalt
var. Tholeiitic basalt

The first and second photographs are of a tholeiite which is somewhat coarser in grain size than is usual. Clinopyroxenes subophitically enclose laths of plagioclase. The brown interstitial regions are of much finer grain size and consist of plagioclase, clinopyroxene, opaque mineral and devitrified glass. A few skeletal oxide crystals are visible.

The third photograph shows an XPL view of an olivine tholeiite. The few rounded crystals showing bright interference colours are microphenocrysts of olivine in a groundmass of clinopyroxene, plagioclase and interstitial glass.

First and second photographs: Tholeiite from Deep Sea Drilling Project from Leg 34; Nacza Plate S.E. Pacific magnification × 11, PPL and XPL.
Third photograph: Olivine tholeiite from Columbia River, USA; magnification × 20, XPL.

119

Basalt

var. Alkali olivine basalt

The photographs show a rock consisting of brownish augite crystals subophitically enclosing clear laths of feldspar. One microphenocryst of plagioclase feldspar is clearly visible at the top edge of the field of view. Two microphenocrysts of olivine are easily identified by their bright interference colours – one blue crystal adjacent to the feldspar microphenocryst at the top edge of the photograph and one pink crystal to the right of the centre of the field. Smaller crystals of olivine can be identified by comparing the two photographs – in the PPL view the olivine crystals have much paler colour than the clinopyroxenes, which in this rock are quite strongly coloured.

Alkali olivine basalt from Hawaii; magnification × 15, PPL and XPL.

120

Basalt

Lunar low-Ti basalt

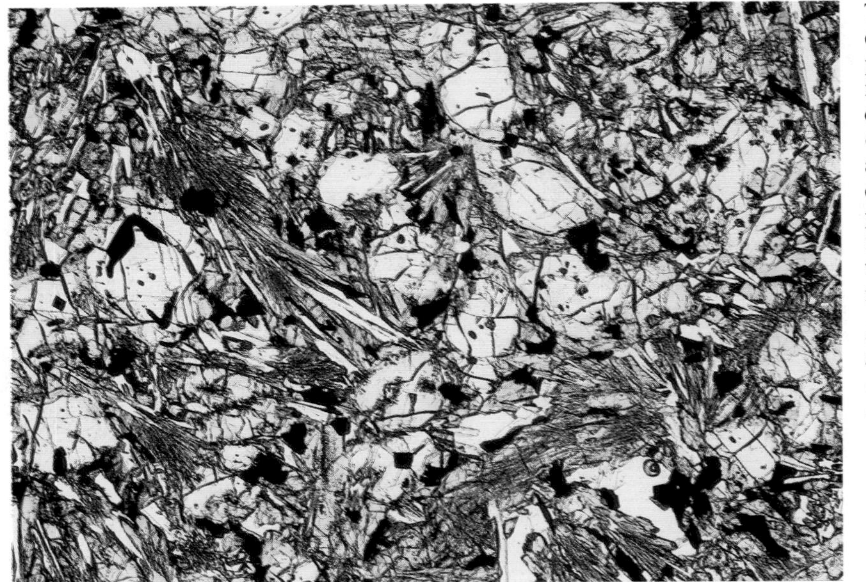

The photographs show phenocrysts of olivine and pyroxene set in a matrix of plagioclase and, pyroxene. The pyroxenes can be distinguished from the olivines by the fact that they have a reddish-brown colour, the intensity of which increases towards the rims of the crystals: the olivines are nearly colourless in PPL and within them are inclusions of glass and also opaque minerals, mainly chromite and an Fe-Ni alloy. There is rather a small proportion of plagioclase in the view shown here; it forms elongated crystals intergrown in variolitic fashion with pyroxene. The opaque mineral in the groundmass is mainly ilmenite.

Low-titanium porphyritic basalt obtained by the Apollo 12 mission from the Riphaeus Mountains, south of Copernicus (NASA sample number 12002; 399); magnification ×28, PPL and XPL.

121

Basalt

Lunar high-Ti basalt

The thin section of this rock shows reddish-brown pyroxene, plagioclase and ilmenite. The completely unaltered state of the pyroxene and plagioclase is probably the first characteristic which strikes the observer. Zoning in the pyroxene is distinct in some crystals. Notice the unusual texture in which plagioclase laths subophitically enclose pyroxene, contrary to normal subophitic texture. Accessory cristobalite is also present. It can best be seen in the top right-hand corner of the view in PPL, where its low refractive index means that it stands out in relief against the calcic plagioclase – its low interference colours are seen in the view under crossed polars.

Coarse-grained high-titanium basalt obtained by the Apollo 17 mission from the Taurus-Littrow Valley, Taurus Mountains. (NASA sample number 70017, 216); magnification ×25, PPL and XPL.

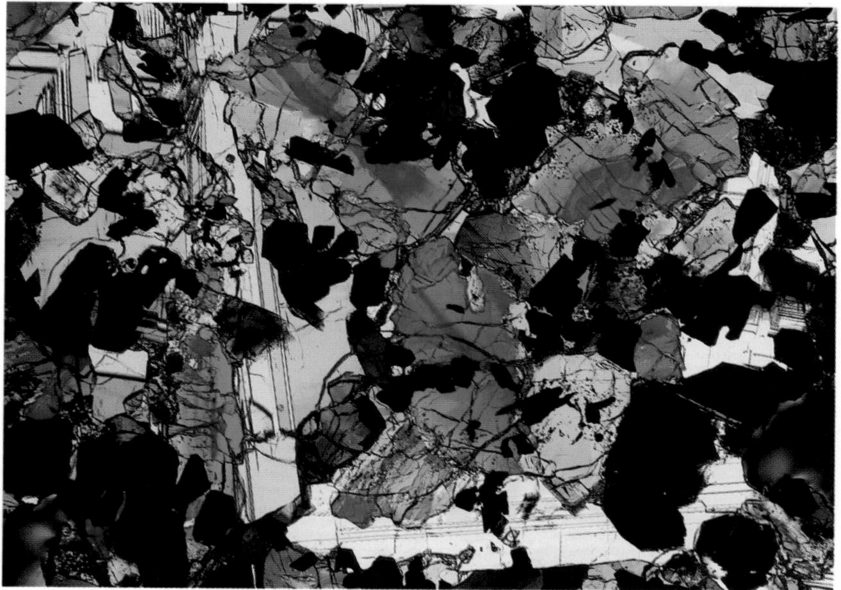

122

Basalt

var. Picritic basalt

This sample has abundant phenocrysts and glomerocrysts of subhedral and euhedral olivine in a groundmass of olivine, pyroxene and small proportions of plagioclase and iron ore.

Picritic basalt from Ubekendt Island, W. Greenland; magnification × 8, PPL and XPL.
See also ***26**, **27** and **31**.*

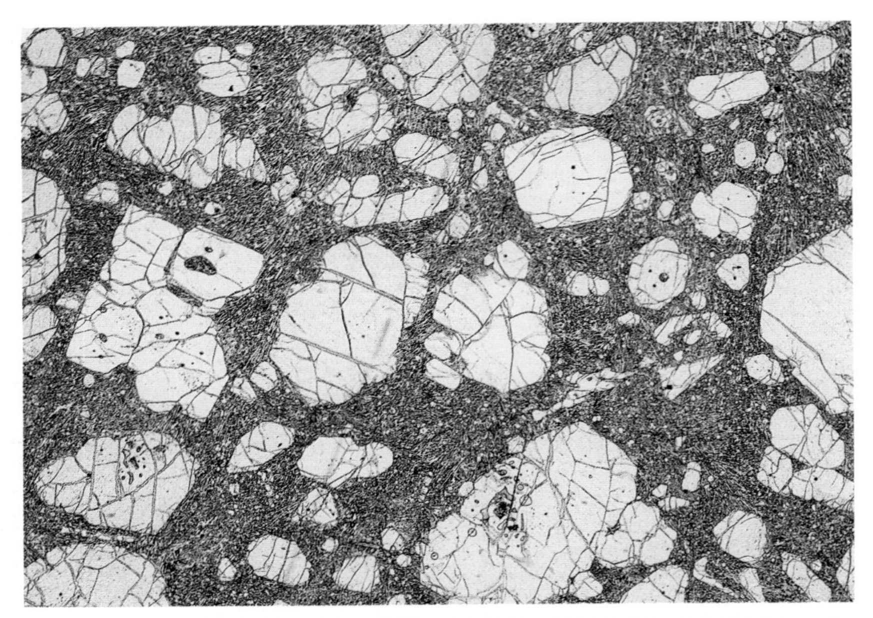

123

Basalt

var. Ankaramite

The specimen illustrated here has zoned phenocrysts and microphenocrysts which are mainly of pyroxene, although a few olivine crystals can be seen. The olivine crystals occur mostly as microphenocrysts and they are slightly paler in colour in the PPL photograph than the pyroxenes. We can identify two of these olivines in the field of view – one shows a blue interference colour and is just to the right of centre at the top edge of the photograph, and another is at the centre of the bottom edge of the photograph and shows a pale yellow interference colour. The groundmass contains minute laths of plagioclase, too small to be visible at this magnification, embedded in cryptocrystalline material.

Ankaramite from Ivohitra, Madagascar; magnification ×15, PPL and XPL.

124
Spilite

A basic rock, commonly amygdaloidal, in which the original minerals have been affected by some type of alteration, so that the feldspars are of albite composition and the pyroxenes have been replaced by other minerals. Although it has been suggested many times that this name should be dropped on the grounds that spilites are metamorphosed basalts it is still in use.

The sample illustrated shows an amygdaloidal rock in which the amygdales are filled with calcite, white in both the XPL and PPL views and chlorite, green in the PPL view and showing anomalous blues and purples in the XPL view. In the groundmass of the rock are laths of feldspar of albite composition set in dark patches which are mainly of finely crystallized chlorite, calcite and haematite.

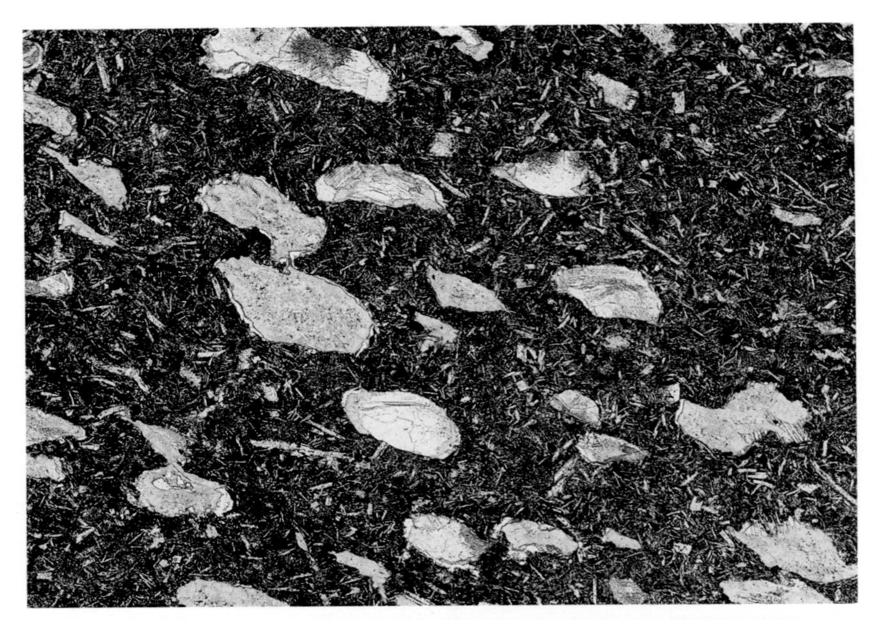

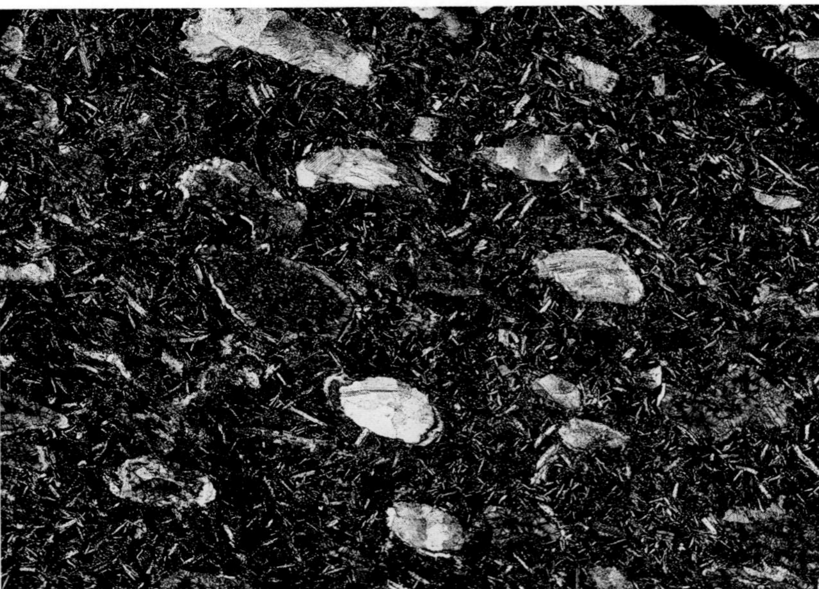

Spilite from Chipley quarry, Devon, England; magnification ×43, PPL and XPL.

125
Gabbro

A coarse-grained, dark- to medium-coloured rock consisting essentially of augite and a plagioclase of labradorite, or more calcic, composition, i.e. the equivalent of basalt and dolerite. Other minerals which may be present are orthopyroxene, pigeonite, olivine or quartz. Gabbros of tholeiitic affinity commonly show lamellar textures in the pyroxenes.

Gabbros containing feldspathoids (including analcite) and lacking low-Ca pyroxene have the group name *alkali-gabbros* or *syenogabbros*. A number of names are used for these alkali gabbros: if the gabbro contains abundant analcite and little nepheline *teschenite* (**126**) is used; if nepheline amounts to more than 10% *theralite* is used, or if olivine is present it is termed an *olivine theralite*; if nepheline exceeds 10% and between 10% and 40% of the feldspar is of alkali type, the name *essexite* (**127**) is used.

The rock illustrated is a granular olivine gabbro. A group of three olivine crystals are located at the centre of the field towards the top and another crystal is at the right edge of the field. The rest of the field is occupied by augite and a basic plagioclase – the section is slightly too thick since some of the plagioclase crystals show a very pale yellow interference colour. The augite shows slight zoning in some crystals and a lamellar texture is also visible; this is probably due to exsolution. Absence of orthopyroxene and presence of olivine make this rock the coarse-grained equivalent of olivine tholeiite.

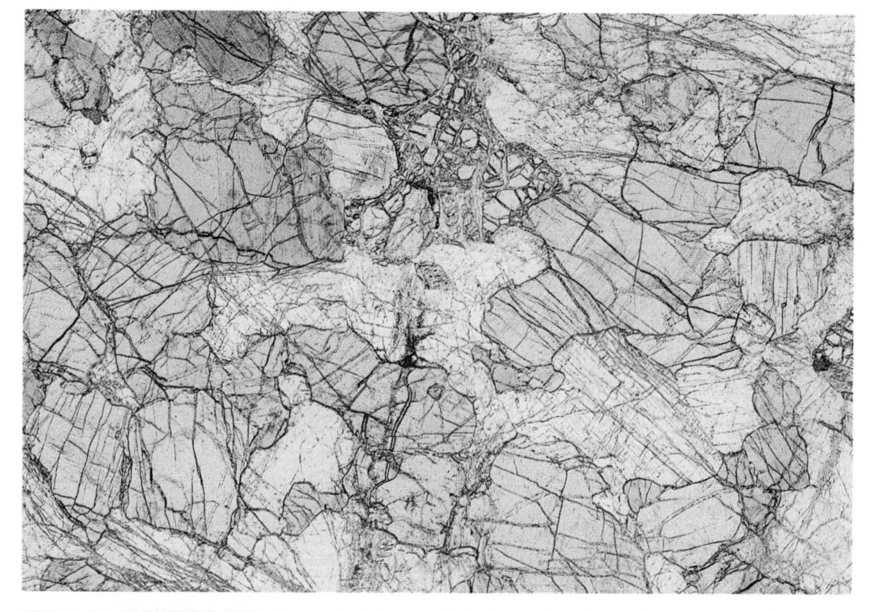

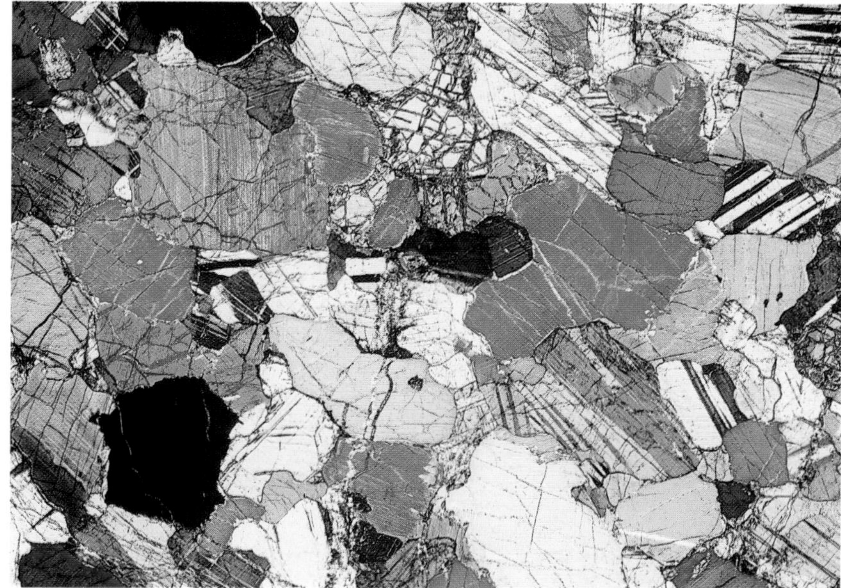

Olivine gabbro from New Caledonia; magnification × 11, PPL and XPL.
*Other gabbros are illustrated in **15, 16, 17, 40, 49, 50, 61, 68, 93**.*

126
Teschenite

This is the name used for an alkali gabbro or dolerite which consists essentially of a plagioclase feldspar of labradorite or more calcic composition, a clinopyroxene which is commonly a purplish-brown colour, and analcite. If olivine is present, *olivine teschenite* is used. The term *crinanite* is sometimes used to denote a medium-grained olivine-bearing rock of this type which contains only trivial, but essential analcite.

The sample we have illustrated contains olivine, purplish clinopyroxene, plagioclase and analcite. Olivines, ranging in size from 0.2–2 mm, are scattered throughout the rock and may be recognized by their grey colour in PPL. The subophitic clinopyroxene shows the colour typical of titaniferous pyroxenes and zoning can be clearly seen in the large crystal near the top of the field just to the left of centre.

In the XPL view, the analyzer has been rotated through a few degrees so that the analcite can be distinguished from the opaque ores present in the rock – instead of being completely black the analcite has a slightly brownish colour (e.g. left of centre); the triangular crystal of iron oxide near the top centre of the field is partially surrounded by analcite. This rock also contains some nepheline but it cannot be easily seen in this photograph. Note the unusual radiate arrangement of plagioclases at lower centre.

Teschenite from Dippin sill, Arran, Scotland; magnification ×5, PPL and XPL.

127

Essexite

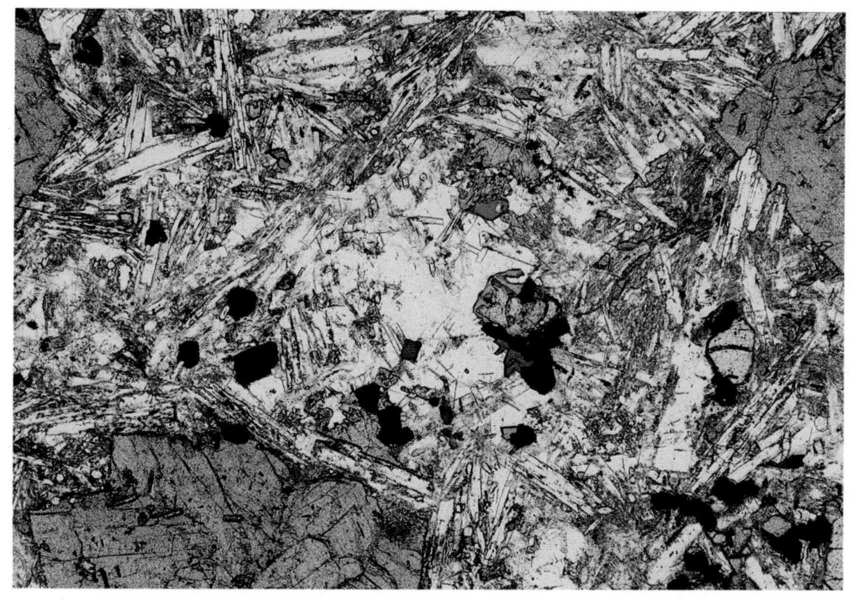

This name is used for a medium- or coarse-grained rock consisting essentially of labradorite or a more calcic feldspar, clinopyroxene and olivine with small and variable minor amounts of alkali feldspar, and a feldspathoid which is nepheline with or without analcite (see **125**).

The upper photograph shows an XPL view of a rock consisting of large zoned clinopyroxene phenocrysts, small olivine crystals, laths of plagioclase, iron ore and analcite. An enlarged view of the central area of this photograph is produced as the middle and lower photographs. In the PPL view, the brown colour of the clinopyroxene crystals serves to distinguish them from the olivines – there is one olivine crystal fairly close to the right edge of the photograph and one just to the right of the centre of the field, partly surrounded by biotite and partly by iron ore. Much of the large clear area in the centre of the PPL view is analcite but within this area there are alkali feldspar and nepheline crystals whose relief is such that they do not stand out against the analcite as clearly as does the calcic plagioclase. The small hexagonal and needle-shaped crystals of high relief are of apatite which is widely distributed in this rock. One useful observation is that alkali feldspar commonly rims plagioclase crystals, as can be seen in the long plagioclase crystal lying sub-parallel to the right edge of the enlarged XPL and PPL photographs; at the lower extremity of this crystal there is a rim of alkali feldspar.

Essexite from Crawfordjohn, Scotland; magnification × 7 (upper), XPL, × 26 (middle and lower), PPL and XPL.

128
Dolerite

This is the name used for medium-grained basic rocks consisting essentially of labradorite, augite and ore minerals, i.e. equivalent of basalt and gabbro. In North America the term *diabase* is used in preference to dolerite to denote the same rock. Like basalts and gabbros, there are tholeiitic and alkalic varieties which can be identified from the presence or absence of low-Ca pyroxenes, nepheline, analcite, quartz and the absence or presence and amount of olivine. Coarse-grained names are often prefixed by *micro-* to name alkalic varieties (e.g. micro-teschenite).

The photographs show PPL and XPL views of a sub-ophitic tholeiitic dolerite – this is confirmed by the presence of both orthopyroxene and clinopyroxene. A small amount of olivine is present in this rock and it has been partly replaced by serpentine – this can best be seen in the PPL view where the olive-green patches are of serpentine after olivine. Most of the pyroxene at the lower left of the field of view is orthopyroxene, it has a lamellar texture but this cannot be seen at this magnification. The crystals showing red and blue interference colours to the right of the field of view are of clinopyroxene. The ophitic texture is fairly typical of this type of rock.

Dolerite from Palisades sill, New York, USA; magnification ×21, PPL and XPL.
Additional dolerites are illustrated in ***52, 53, 59, 60, 63, 89*** *and* ***90****.*

129

Norite

This is the name used for a coarse-grained rock consisting mainly of a calcic plagioclase and orthopyroxene.

The sample illustrated shows plagioclase, orthopyroxene and some clinopyroxene in a subhedral granular texture. In the PPL view it is difficult to distinguish the two pyroxenes but we can detect the presence of intergrowths in the two largest areas of brownish pyroxene. In the XPL view the areas with a brownish-yellow interference colour are of orthopyroxene with clinopyroxene lamellae showing higher interference colours. Smaller crystals showing blue and green interference colours are of clinopyroxene and these have lamellae of orthopyroxene. One such area showing a blue interference colour is located about the centre just above the bottom edge of the field.

Norite from Bushveldt complex, South Africa; magnification × 12, PPL and XPL.

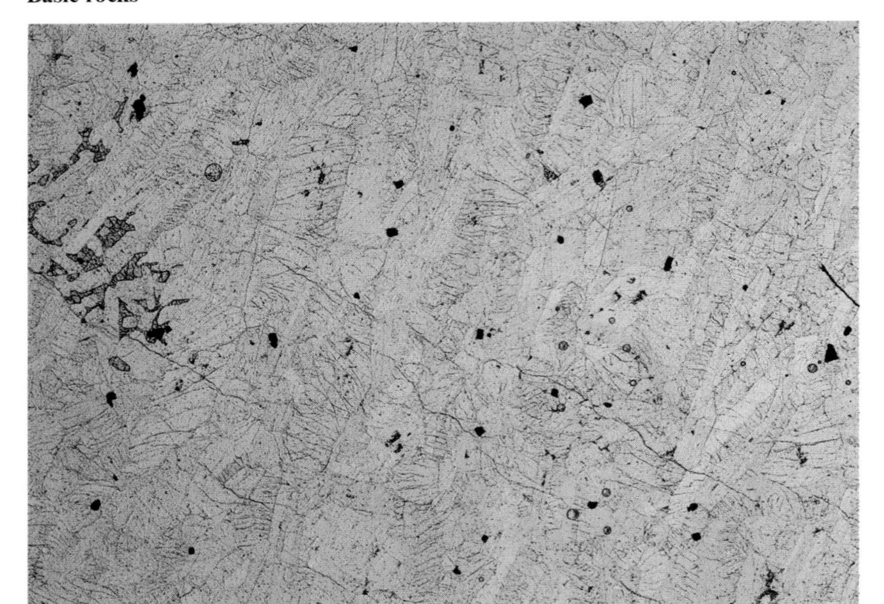

130

Anorthosite

An anorthosite is a coarse-grained rock consisting of more than 90% plagioclase, usually labradorite or bytownite. There are a variety of different types of anorthosite depending on the type of occurrence and the composition of the plagioclase; many anorthosites, although perhaps originally igenous, are now metamorphic rocks.

The specimen illustrated in the upper and middle photographs is of an anorthosite layer from an igneous complex. Most of the field of view is occupied by bytownite; towards the left edge of the field of view some pyroxene can be recognized among the plagioclase crystals. The mineral which appears opaque is a very dark brown chrome spinel. There is a strong preferred orientation in the tabular plagioclase crystals in this rock giving it a laminated texture. The thin section is slightly thick as is indicated by the pale yellow tinge in the plagioclase interference colours.

The lower photograph is of a lunar anorthosite. It shows a strained and broken (i.e. cataclastic) texture which probably resulted from meteorite impact. The feldspar composition in this sample is An_{97} and is much more calcium-rich than most terrestrial anorthosites. Only tiny crystals of pyroxene exist among the small feldspar fragments. The large feldspar crystal occupying the lower left part of the field of view shows patchy extinction as does the crystal just above the centre of the field. This material is one of the oldest known rocks having an age between 4,000 and 4,500 million years.

First and second photographs: Anorthosite from Rhum, Scotland; magnification × 9, PPL and XPL.
Third photograph: Anorthosite from Cayley Formation, Descartes region of the Lunar Highlands; sample brought back by Apollo 16 mission (NASA sample number 60025, 255); magnification × 12, XPL.

*Another anorthosite is illustrated in **104**.*

131

Andesite

This is the name given to a volcanic rock with essential andesine in the groundmass and one or more ferromagnesian minerals, commonly pyroxene(s) or hornblende ± biotite. The rock type is commonly porphyritic and the feldspar phenocrysts may be complexly zoned and embayed with the composition of the cores of the crystals as calcic as bytownite. The equivalent coarse-grained rock is *diorite*.

The first photograph shows a PPL view of a porphyritic rock in which the phenocrysts are of plagioclase, a brown amphibole and a few scattered microphenocrysts of pyroxene in a microcrystalline groundmass consisting mainly of plagioclase. Some of the plagioclase phenocrysts are quite different from others: at the right edge of the field, a large square phenocryst contains numerous inclusions and has a dark rim of inclusions.

The second and third photographs show an andesite containing clear phenocrysts of plagioclase and brownish phenocrysts of pyroxene. The view has been selected to include orthopyroxene and augite in the group of crystals to the right of the centre of the field. The lower crystal, lighter in colour in the PPL view, is orthopyroxene, the darker crystal is augite. The groundmass contains plagioclase and pyroxene but this is too fine grained to determine the nature of the pyroxene.

First photograph: Andesite from Bolivia; magnification × 17, PPL.
Second and third photographs: Two-pyroxene andesite from Hakone volcano, Japan; magnification × 9, PPL and XPL.

*Additional andesite photographs are shown in **19** and **97**.*

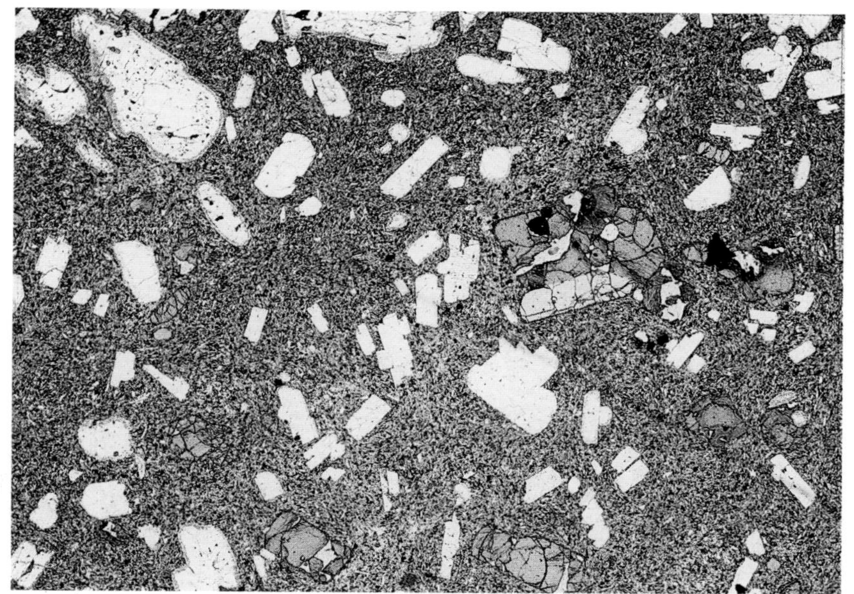

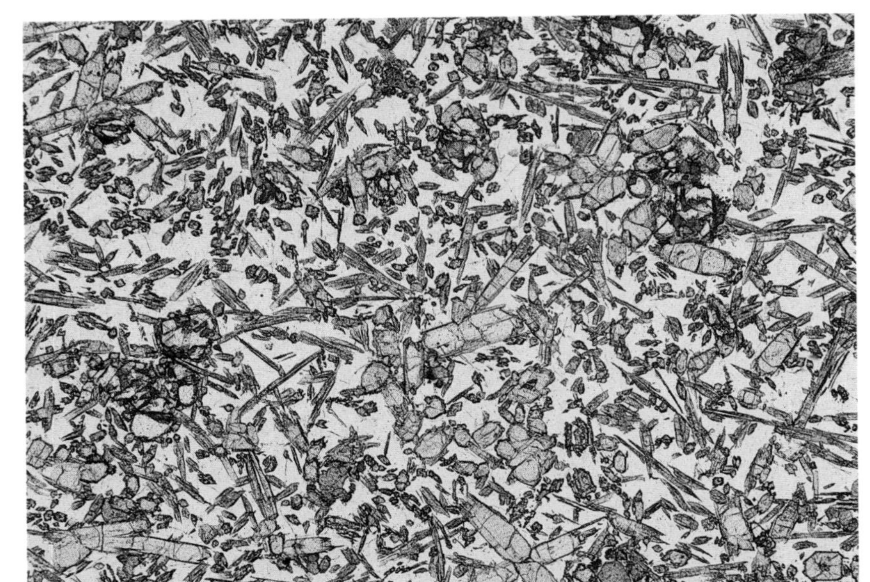

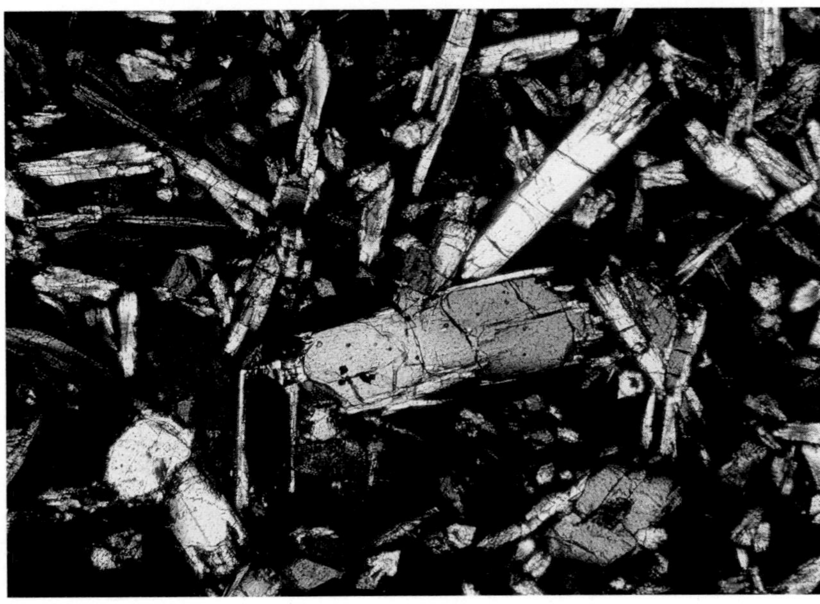

132
Boninite

This is the name given to a volcanic rock the composition of which is that of a high magnesia andesite. It is predominantly glassy but contains microphenocrysts of orthopyroxene and sometimes olivine and clinopyroxene also.

The specimen illustrated shows orthopyroxene crystals with first-order grey interference colours, clinopyroxenes showing first- and second-order colours and somewhat larger skeletal crystals of olivine. Alteration of the olivines to serpentine is seen in brownish patches in the thin section. No plagioclase is visible in any of these photographs.

The third photograph is an enlargement of the central area of the other photographs. This shows a large orthopyroxene crystal with two clinopyroxene crystals parallel to the length of the orthopyroxene. At the left extremity of this orthopyroxene is another one (at extinction) mantled on three sides by clinopyroxene.

Boninite from Chichi Jima, Bonin Islands, Japan; magnification × 23 (first and second photos), PPL and XPL, × 72 (third photo), XPL.

133
Diorite

A coarse-grained, medium-coloured rock consisting of andesine plagioclase and one or more of the mafic minerals clino- and ortho-pyroxene, hornblende and biotite. A little quartz and K-feldspar may be present. Olivine is a minor constituent of some diorites. This is the coarse-grained equivalent of andesite. If quartz is abundant (up to 20%) the name *quartz diorite* is used.

The photographs show a field of view in which there are three mafic minerals: biotite, pale to dark brown colours in the view under PPL; orthopyroxene; and clino-pyroxene. Some of the clinopyroxene crystals can be easily recognized in the XPL view because they show second-order interference colours of blue and purple. The orthopyroxenes show only first order colours. The plagioclase feldspar shows a very slight clouding which can be detected in the PPL view. A rock from the same intrusion showing pronounced clouding of the feldspar is shown in Part I in **67**. A small amount of quartz is present in this rock and is recognized by absence of alteration.

Diorite from Comrie, Perthshire, Scotland; magnification × 12. PPL and XPL.

Another diorite is shown in ***92***.

134
Tonalite

A coarse-grained rock consisting of a plagioclase (approx. An_{30}), hornblende, or biotite or both, and accessory quartz. The name is synonymous with *quartz diorite*. (In North America tonalite implies a quartz content of more than 20%, but in Britain smaller quartz contents are included.)

The mineralogy of the rock illustrated is simple: zoned plagioclase and strained quartz are readily identified, and the only ferromagnesian mineral present is biotite. The cores of the plagioclase crystals contain a dense mass of very fine mica and clay minerals, generally ascribed to alteration, and the biotite crystals have inclusions, many of which produce pleochroic haloes.

Tonalite from Theix, Puy de Dôme, France; magnification × 12, PPL and XPL.

135
Kentallenite

A coarse-grained rock consisting of essential olivine, augite, biotite, plagioclase (more calcic than An_{50}) and orthoclase. This is a coarse-grained equivalent of trachybasalt (i.e. a syenogabbro), it may be regarded as an olivine-bearing monzonite.

In the porphyritic specimen illustrated here, the ferromagnesian minerals olivine, augite and biotite are fairly easily identified in the view in PPL. Most of the biotite crystals show a fox-brown absorption colour. The crystals covered by networks of black cracks are olivines, and the augite crystals can be identified by their relief against the colourless feldspar: there are only three augite crystals visible and two of these show yellow interference colours and the third, at the top right of the field of view, shows a blue interference colour. Some pale green chlorite can be seen near the left edge of the photographs.

The mineral showing grey, black and white interference colours is largely labradorite. It is difficult to show that orthoclase is present in this field of view but two untwinned crystals with uniform grey interference colours, just below the olivine crystal showing a purple interference colour at the right edge of the field, are orthoclase.

Kentallenite from Kentallen, Scotland; magnification × 12, PPL and XPL.

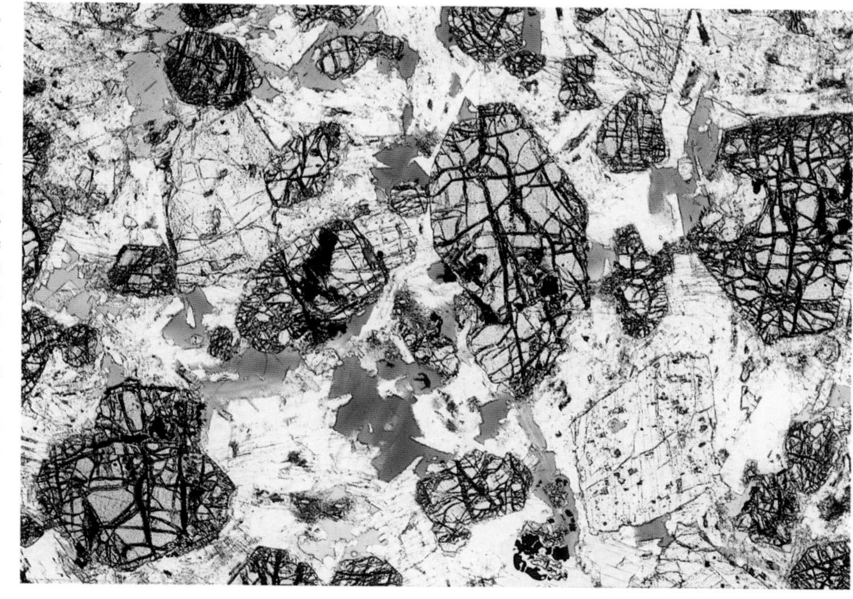

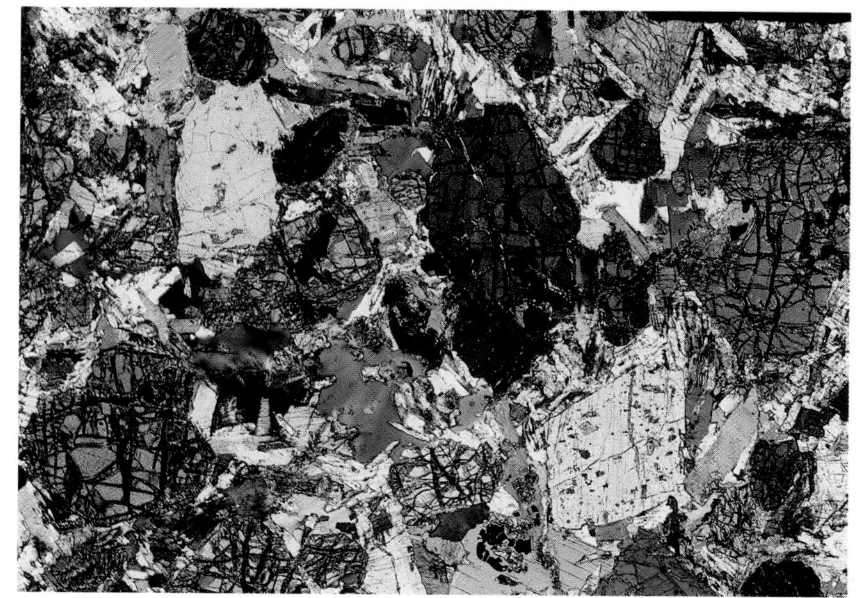

136
Monzonite

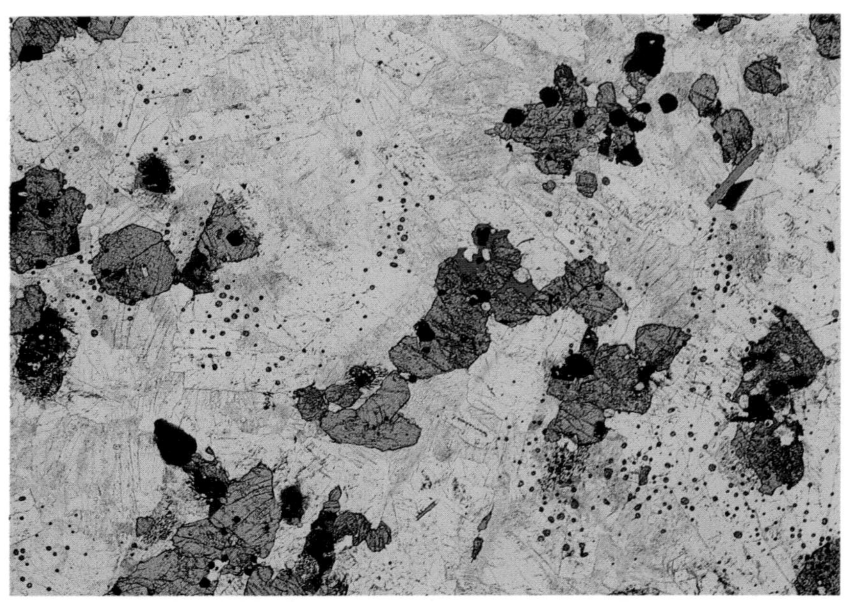

This is the name used for a coarse-grained rock in which the amount of alkali feldspar is about the same as that of plagioclase feldspar. The amount of quartz present should be 0–5%, and mafic minerals 10–25%. The rock is intermediate in character between *syenite* and *diorite*. Its fine-grained equivalent is called *latite* or *trachyandesite*.

The rock illustrated consists mainly of clinopyroxene, alkali feldspar and plagioclase feldspar. In the PPL view the clinopyroxene can be readily identified, as can small amounts of biotite and opaque oxides. In this rock the distinction between plagioclase and alkali feldspar is easily made because the alkali feldspar shows no multiple twinning, whereas almost all the plagioclase crystals in the field of view are multiple twinned. Very fine intergrowths are present in the alkali feldspar but they cannot easily be seen at the magnification of these photographs. A very small amount of quartz is present in this rock.

Monzonite from Mount Dromedary Complex, New South Wales, Australia; magnification × 11, PPL and XPL.

137

Dacite

This is a volcanic rock which usually contains phenocrysts of plagioclase, less calcic than andesine, quartz and subordinate ferromagnesian mineral(s) in a fine-grained groundmass: a small proportion of alkali feldspar may be present. The coarse-grained equivalent is *granodiorite*.

The specimen illustrated shows strongly-zoned phenocrysts of plagioclase, and quartz and microphenocrysts of a brownish-green amphibole in a fine-grained groundmass composed of the same minerals and biotite. The phenocrysts of quartz can be distinguished from those of plagioclase by a number of criteria: the interference colour shown by the quartz crystals is just slightly higher than that of the feldspar; the quartz crystals show no zoning or twinning; and they have round outlines, whereas the plagioclase crystals show the outlines of crystal faces only slightly rounded at the corners. Some of the plagioclase phenocrysts have sieve texture (top left). The feldspars show a complete gradation in size of crystals from phenocrysts through microphenocrysts to groundmass crystals (seriate texture). Some of the variation in size of the crystals is due to the fact that the section does not cut through the centres of all crystals.

Dacite from unknown locality in Argentina; magnification ×8, PPL and XPL.

Another dacite is illustrated in ***3****.*

138
Granodiorite

A granodiorite is a medium- to light-coloured coarse-grained rock containing essential quartz, plagioclase feldspar (oligoclase), and alkali feldspar, in amounts between 10% and 35% of the total feldspar, and lesser amounts of mafic minerals, commonly hornblende or biotite, or both. Whereas granodiorites have 20–30% quartz, diorite has only accessory quartz, if any.

The specimen illustrated is a granular biotite-hornblende granodiorite. The hornblende is pleochroic in shades of brown and green in PPL view (e.g. the three green crystals near the centre of the field). Two of these green hornblende crystals show twinning in the XPL view. Just below them are two biotite crystals showing light and dark brown absorption colours. Quartz can be distinguished from the feldspar in that it is relatively clear in the PPL view and some crystals show a slightly higher interference colour than the feldspars. It is more difficult to distinguish the alkali feldspar from the plagioclase, but in this rock the former shows only simple twins: the crystals at the top right corner of the field of view are of alkali feldspar, whereas the large crystal near the centre of the field is plagioclase. The distinction is most easily made by the difference in relief which cannot easily be shown in photographs, but the plagioclase in this rock has a higher relief than quartz whereas the alkali feldspar has a lower relief.

Granodiorite from Criffel-Dalbeattie, Scotland; magnification × 11, PPL and XPL.

139

Trachyte

A trachyte is a fine-grained volcanic rock consisting mainly of alkali feldspar or sometimes of two feldspars, a sodic plagioclase and a potassic feldspar. A small amount of ferromagnesian minerals is usually present. Quartz or nepheline may be present as accessories but are confined to the groundmass. The coarse-grained equivalent is *syenite*.

The sample illustrated is a porphyritic rock in which the predominantly euhedral phenocrysts are of both sodic plagioclase and sanidine. In the centre of the field there is a group of plagioclase crystals which in the PPL view can be seen to have slightly higher relief than the large sanidines which occupy the bottom left corner of the field. In the XPL view the multiple twinning in the plagioclase crystals is obvious. The two large crystals at the right of the photograph are of sanidine, one showing a simple twin. At the bottom left there is a hole in the slide in which the large sanidine shows broken fragments. A few microphenocrysts of pyroxene are visible, one lying just above the centre of the field. Unfortunately it does not show up very well in the PPL view because the substage diaphragm was stopped down to show the relief in the plagioclase feldspar, and in the XPL view this pyroxene is in the extinction position.

Trachyte from Ischia, Italy; magnification × 14, PPL and XPL.

Additional trachyte photographs are shown in ***64, 65*** *and* ***107****.*

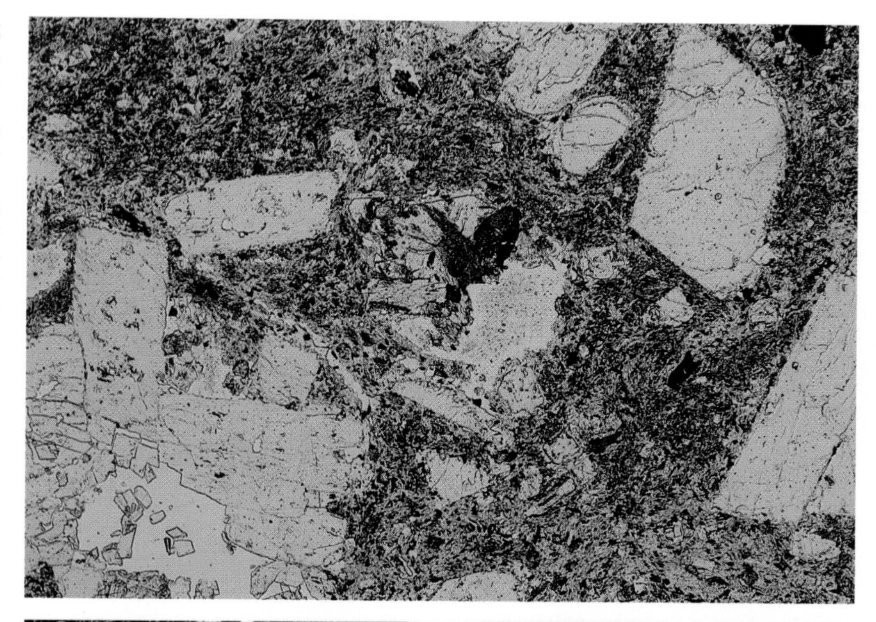

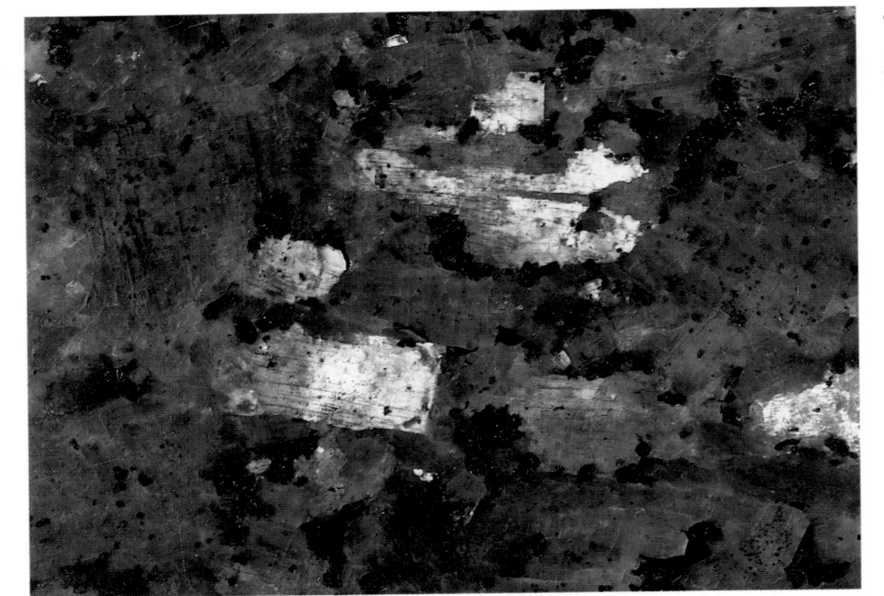

140
Syenite

A syenite is a light-coloured, coarse-grained rock consisting mainly of alkali feldspar with less than 5% quartz, or less than 5% of feldspathoid. Clinopyroxene, hornblende, biotite and even olivine may be present in small amounts. Plagioclase may be present as an accessory. This rock is the plutonic equivalent of *trachyte*.

The upper photograph is of a polished slab of *larvikite*, a syenite commonly used as a decorative building material because of the moonstone schiller shown by the alkali feldspars. A thin section cut from this rock is illustrated in XPL view in the middle photograph. Most of the field of view is occupied by cryptoperthitic alkali feldspar. A small amount of nepheline is present in this sample but cannot be easily illustrated in one photograph. The crystal showing a green interference colour at the right edges of the field is an iron-rich olivine: clinopyroxene and biotite are also present.

The lower photograph is of a granular-textured syenite. A field of view has been chosen so as to show the ferromagnesian minerals at the edges only. The mineral on the top edge showing a brown interference colour is of pyroxene; a purplish absorption colour being added to the interference colour is the cause of the slightly anomalous colour. The crystals showing red and green colours at the right-hand edge are olivines with alkaline amphibole rims to the crystals.

First and second photographs: Syenite from Larvik in Norway; magnification × 1 (first photo), × 11 (second photo), XPL.
Third photograph: Syenite from Ilimaussaq, South-west Greenland; magnification × 16, XPL.

141
Shonkinite

This name is used for a melanocratic or mesocratic potassic *syenite*.

The photographs show crystals of biotite and clinopyroxene, with one olivine crystal towards the bottom left corner of the field – its interference colour in the XPL view is dark green. Olivine is only an accessory constituent of a *shonkinite*. The remainder of the field of view is occupied by an alkali feldspar with a symplectite-like intergrowth, best seen in the PPL view; some sodic plagioclase is also present. The parts of the symplectite which appear brown are fine grained alteration products of another mineral. A small amount of nepheline is also visible in this view but is not easily distinguished in the photograph. With increasing amount of nepheline this rock would grade into a *malignite* (see **152**).

Shonkinite from Shonkin Sag, Highwood Mountains, Montana, USA; magnification × 22, PPL and XPL.

142
Rhyolite

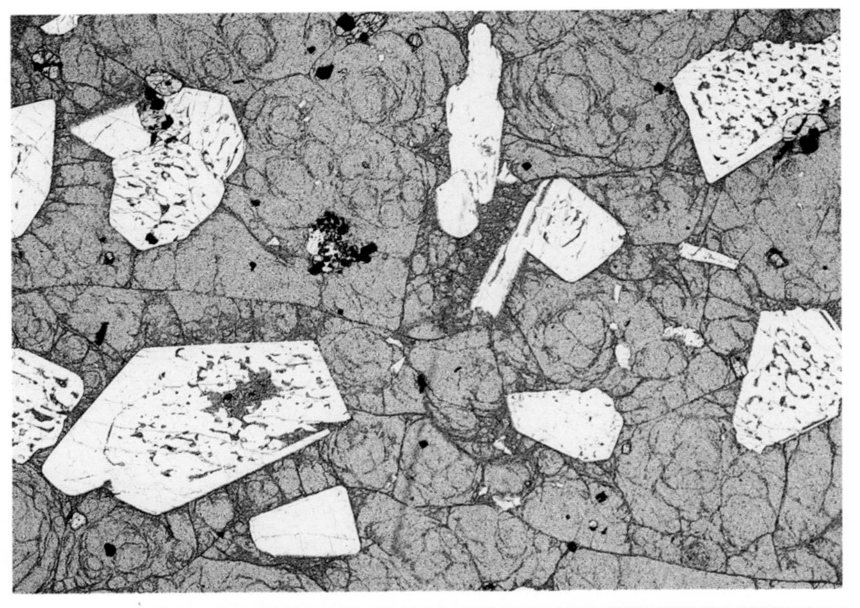

A rhyolite is an acid volcanic rock generally containing phenocrysts of quartz and alkali-rich feldspar in a fine-grained or glassy groundmass. Unfortunately, because the quartz phenocrysts may be absent, it is sometimes impossible to state without a chemical analysis whether a given rock is a rhyolite. The feldspar may be sanidine, sodic plagioclase or both. *Granite* is the coarse-grained equivalent. Many rhyolites are wholly glassy and some have a high proportion of glass – the terms *obsidian* and *pitchstone* then apply respectively; these terms do not, however, imply rhyolite composition. It may be noted that some petrologists define obsidian and pitchstone not on the basis of crystallinity but on water content: in the former it is usually less than 1% and in the latter up to 10%.

We have chosen a rock containing two types of feldspar phenocrysts in a microcrystalline to glassy groundmass showing perlitic fracturing. Most of the phenocrysts are of alkali feldspar, some of which show simple twinning and contain glass inclusions; one phenocryst in the field is a plagioclase. The small crystals showing higher interference colours are clinopyroxene. Some opaque iron ore crystals are also present.

Rhyolite from Eigg, Scotland; magnification × 7, PPL and XPL.

*Additional views of rhyolites and views of pitchstones are shown in **3**, **5**, **12**, **14**, **21**, **66**, **87**, **88**, **91** and **143**.*

143

Pantellerite

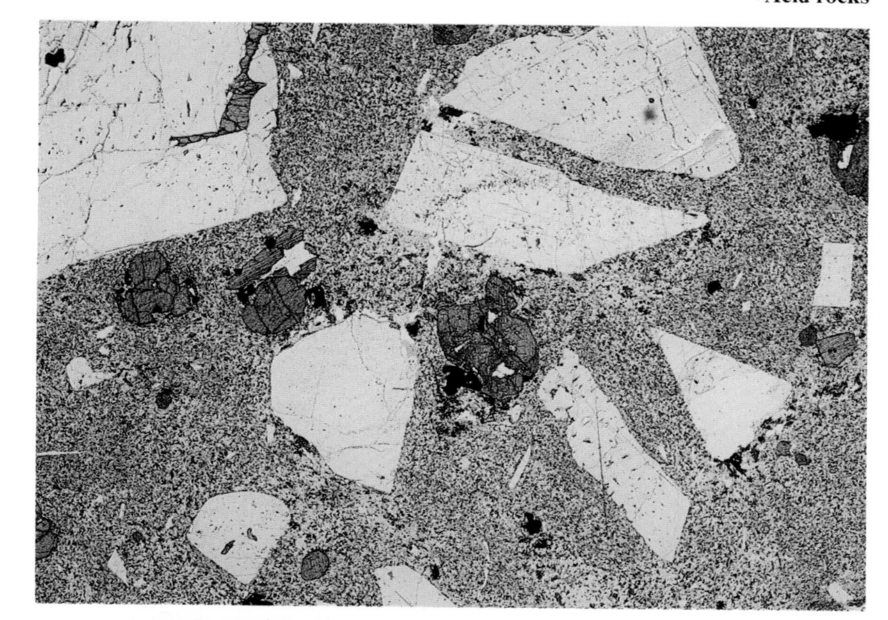

This is the name given to a peralkaline[1] rhyolite. It is usually distinguished from a normal rhyolite by the presence of phenocrysts of anorthoclase and the sodium-titanium mineral aenigmatite. Sometimes an alkali pyroxene is present but it is not always green in colour (see below). Another name used for peralkaline rhyolites is *comendite* but this type is difficult to distinguish from a pantellerite by optical observations only. A comendite tends to have less mafic minerals and is more likely to contain quartz phenocrysts.

We have chosen two samples from the type locality to illustrate this rock. The first and second photographs show phenocrysts of anorthoclase, easily identified by the very fine cross-hatched twinning, together with microphenocrysts of pyroxene and a fayalitic olivine set in a fine-grained groundmass which is mainly quartz and feldspar. Only one olivine crystal is visible in this field of view, it is to the left of centre of the field, showing a blue-green interference colour. The crystals which are almost opaque are of aenigmatite. In the third photograph we have shown a higher magnification XPL view of another sample of pantellerite in which two crystals of aenigmatite showing a dark brown colour are beside a crystal of pyroxene. In these rocks the pyroxene is commonly a sodic hedenbergite, so its absorption colour may be pale brown instead of green.

Two different specimens of pantellerite from Pantelleria, Italy; magnification × 12 (first and second photos), PPL and XPL, × 27 (third photo), XPL.

---

[1] *A rock in which molar* $\frac{Na+K}{Al} > 1$ *(i.e. in which there is a deficiency of alumina with respect to alkalis). This results in crystallization of some alkali pyroxene or alkali amphibole, and, in some cases, aenigmatite.*

144
Granite

Granite is the name used for leucocratic coarse-grained rocks containing mainly quartz and feldspar, alkali feldspar constituting between 90% and 35% of the total feldspar. Accessory hornblende or biotite are the commonest mafic minerals. Muscovite may be present. *Rhyolite* is the fine-grained equivalent. A microgranite with micrographic quartz-alkali feldspar intergrowth is known as a *granophyre* (**77**). A leucocratic microgranite occurring as dykes or veins is known as *aplite*. Granites in which more than 90% of the feldspar is an alkali feldspar are known as *alkali granites* (see **145**). When alkali feldspar is between 35% and 65% of the total feldspar the names *adamellite* and *quartz monzonite* have been used for rocks with quartz content between 5% and 20%.

The first and second photographs are of the Westerly granite, a rock which has been used as a standard granite for a variety of studies. From the photographs it can be seen to consist mainly of a granular mixture of quartz and feldspar, with biotite and a few crystals of muscovite. Sodic plagioclase and potassic feldspar are present: more than half of the feldspar in this rock is potassic feldspar. Much of it does not show microcline-type twinning which appears only in patches in some of the crystals, e.g. just above the centre of the field and slightly to the right is a crystal showing vague cross-hatched twinning. At the left edge of the photograph one potassium feldspar crystal shows a simple twin.

The third photograph is an XPL view of the granite from Shap. In this view the right lower part of the field is occupied by a group of fairly large phenocrysts of alkali feldspar showing a microperthitic texture. Most of the plagioclase crystals show some alteration and can be recognized in the photograph by the presence of multiple twinning and zoning. The quartz crystals can be recognized as free from alteration and the two areas which appear black in this photograph, to the left of the centre and at the left bottom corner, are quartz crystals at extinction. The only other mineral present in a significant amount in this field of view is biotite. (See also the hand specimen photograph in **10**.)

First and second photographs: Granite from Westerly, Rhode Island, USA; magnification × 14, PPL and XPL. Third photograph: Granite from Shap, England; magnification × 7, XPL.

144
Granite
(continued)

The granular textured specimen illustrated here shows mainly quartz and two feldspars. The quartz is recognized in the PPL view by the lack of alteration, and in the XPL view, by its interference colours which are slightly higher than that of the feldspar, and by the non-uniform extinction shown by one crystal at the top right of the view. Microcline is clearly identified by the typical cross-hatched twinning and there are slight signs of microperthitic texture also. In this field of view there are only a few plagioclase crystals visible; one, near the top left corner of the field is a simple twinned crystal showing very dark grey interference colours; vague signs of albite twin lamellae can just be seen in this crystal. Another crystal just to the right of the centre of the field and showing a low grey colour is also plagioclase. To the left of the biotite crystal at middle left is a small patch of myrmekite. From the relative proportions of the two feldspars visible in this field of view this sample could be close to the boundary between alkali granite and granite.

Granite from South Dakota, USA; magnification × 12, PPL and XPL.

Additional views of granites are shown in ***2**, **10**, **42**, **76**, **94**, **105** and **109**.*

145
Alkali granite

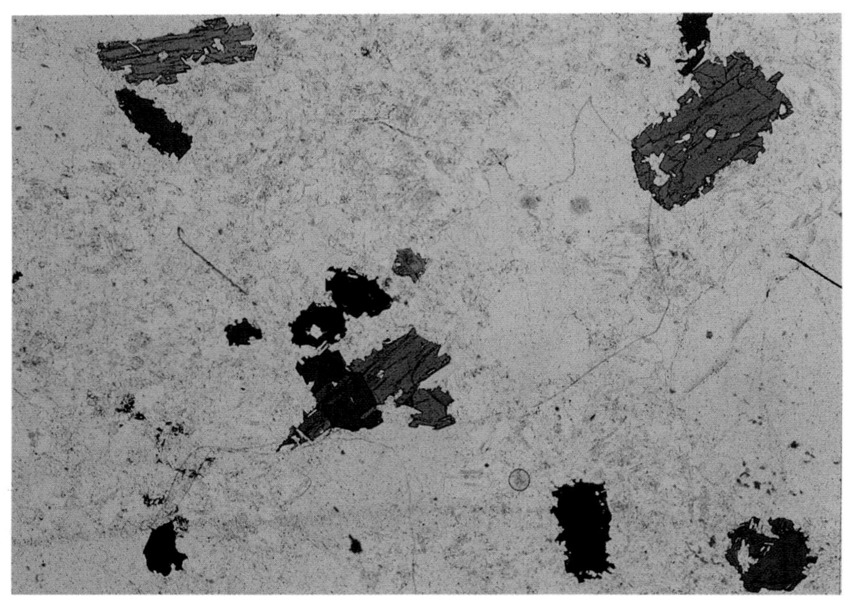

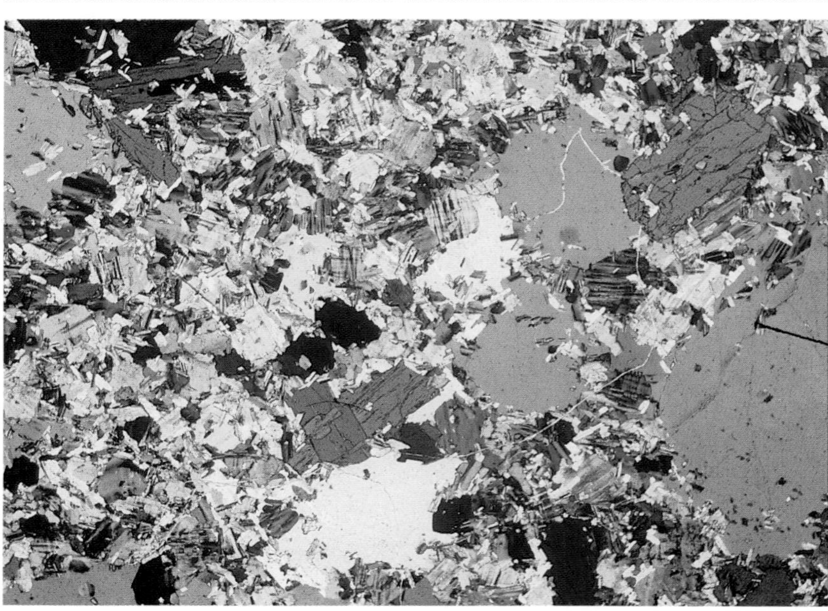

A leucocratic coarse-grained plutonic rock consisting essentially of quartz and alkali feldspar, any plagioclase constituting less than 10% of the total feldspar content. The ferromagnesian minerals present are alkali amphibole, or alkali pyroxene.

The PPL view of this rock shows an area of colourless minerals and a few fairly dark-coloured crystals – these are of an alkaline amphibole, riebeckite, which shows strong pleochroism from a brown to indigo-blue. The interference colours shown in the XPL view are masked by the absorption colours. The large areas of fairly uniform interference colour are quartz phenocrysts, and the rest of the field is made up mainly of albite laths ophitically and subophitically enclosed in subhedral patches of microcline: cross-hatched twinning, characteristic of microcline, is visible. The crystals of albite tend to be smaller than those of microcline and show only albite twinning.

Alkali granite from Jos, Nigeria; magnification × 16, PPL and XPL.

Alkaline and miscellaneous rocks

146
Phonolite

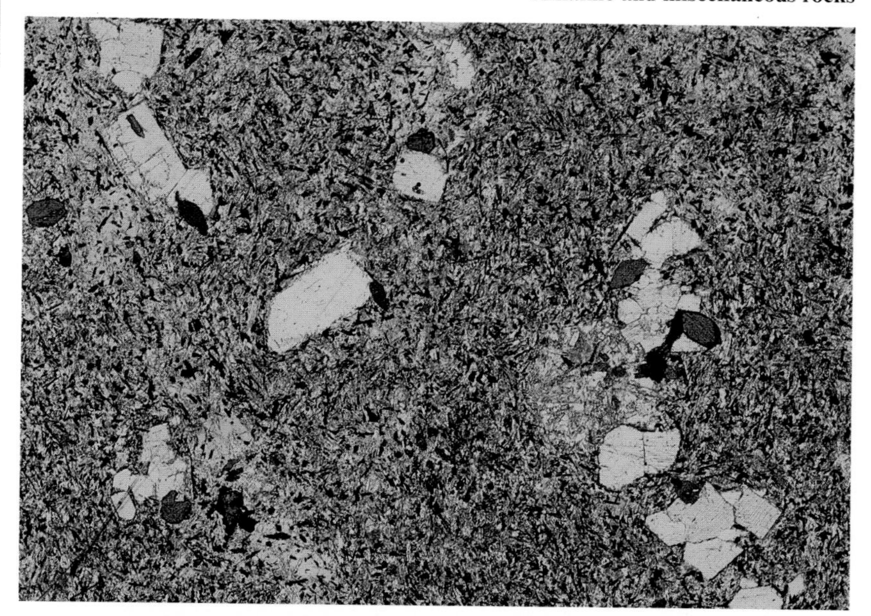

This is the name given to a fine-grained rock consisting chiefly of sanidine or anorthoclase as groundmass laths and frequently also as phenocrysts, nepheline as groundmass crystals and often as phenocrysts as well, with some alkali pyroxene or alkali amphibole. It is fairly common to have a mineral of the sodalite group present also. The coarse equivalent is nepheline syenite.

The first and second photographs show phenocrysts of nepheline and alkali feldspar in a fine-grained groundmass. The greenish-brown microphenocrysts are mainly of alkali amphibole, although in the group of crystals near the bottom left corner of the field there are one or two small crystals of biotite. It is difficult to know which of the phenocrysts are sanidine and which are nepheline but those showing simple twinning are invariably sanidine. Thus the crystal to the left of centre of the field, showing two triangular areas with different interference colours, is almost certainly a Baveno twin of sanidine. In this particular rock the nepheline shows a distinct cleavage and this can be seen in the PPL view in two crystals near to the bottom right-hand corner. This is unusual because, although feldspar may show one or two cleavages, nepheline rarely does so. The groundmass of this rock is made up of sanidine, nepheline and needles of a green pyroxene.

The third photograph is that of a phonolite with a small amount of nosean present. This photograph is included here mainly for historical interest in that the thin section used for this photograph was prepared for J. R. Gregory in 1895 and has been in the Manchester University collection since that time. The view shows nepheline and feldspar phenocrysts and one phenocryst and a few microphenocrysts of nosean, which appear almost black in PPL due to the high density of inclusions. The lath-shaped crystals are likely to be sanidine, whereas the nepheline crystals are rectangular. In this rock the nepheline crystals show zoning by the margins of the crystals having a higher refractive index than the interior, and they can be distinguished from the feldspars by this feature. The groundmass of this rock consists of nepheline, sanidine, a green pyroxene and nosean.

First and second photographs: Phonolite from Marangudzi, Zimbabwe; magnification × 14, PPL and XPL.
Third photograph: Nosean phonolite from Wolf, Rock, Cornwall, England; magnification × 14, PPL.

147
Leucite phonolite

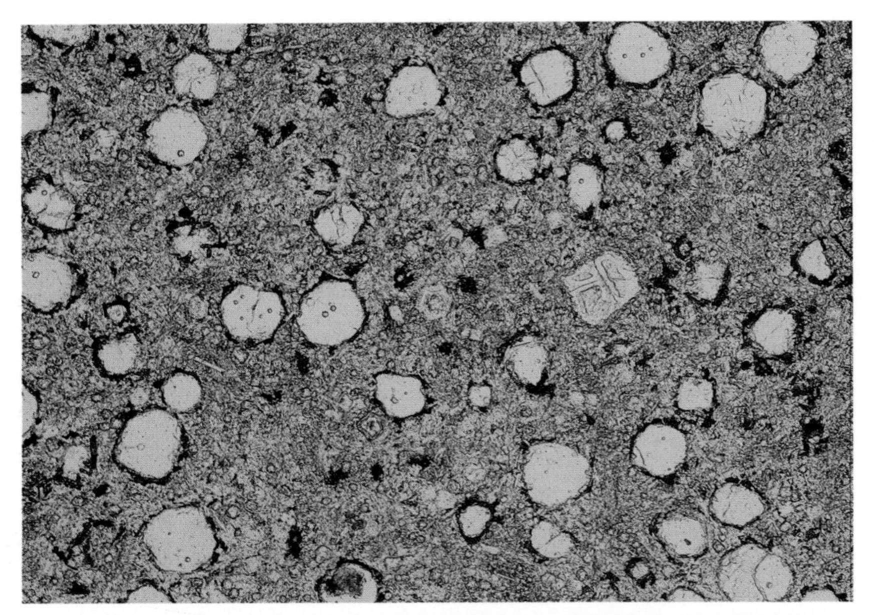

This is the name given to a volcanic rock which consists of essential leucite, nepheline and K-feldspar: an alkali pyroxene is usually present. The term *leucitophyre* was formerly used for varieties which contained no nepheline but this distinction is perhaps not necessary.

The photographs are of a rock which contains leucite microphenocrysts, each of which is surrounded by a ring of small pyroxene crystals. These lie in a groundmass mainly composed of nepheline, leucite, pyroxene and a small amount of alkali feldspar. The nepheline is easily recognized by the rectangular and hexagonal shape of the crystals and one hexagonal cross-section of zoned nepheline is almost exactly in the centre of the field of view.

Leucite phonolite from Olbrück, Eifel, Germany; magnification × 37, PPL and XPL.

148

Nosean leucite phonolite

Most usage of the term phonolite regards nepheline and alkali feldspar as essential constituents so that a nosean leucite phonolite contains nosean, leucite, nepheline and feldspar, usually with an alkali pyroxene as accessory.

The first two photographs show a rock containing phenocrysts of leucite, (clear in the PPL view), brownish nosean crystals with darker brown borders, due to thousands of small inclusions, and phenocrysts and microphenocrysts of a green pyroxene. The multiple twinning in the leucite makes it fairly easy to identify. Around the nosean crystals in the centre of the field the mineral which appears almost white in the XPL view is calcite and it can also be seen fairly well distributed in the groundmass near other nosean crystals. The rest of the groundmass is composed of sanidine, nepheline, nosean, leucite and green pyroxene.

The third photograph shows a very similar rock to the one shown above but the nosean crystals have orange borders. The phenocrysts are of leucite and nosean with microphenocrysts of pale brown pyroxene having greenish rims, and microphenocrysts of leucite and nosean in a groundmass of sanidine, nepheline, nosean, leucite and pyroxene. One elongated sanidine microphenocryst appears at the bottom right of the field. A small amount of sphene and calcite can also be detected in the groundmass but cannot be discerned in the photograph.

First and second photographs: Nosean leucite phonolite from Reiden, Eifel, Germany; magnification × 11, PPL and XPL.
Third photograph: Nosean leucite phonolite from Laacher See, Germany; magnification × 9, PPL.

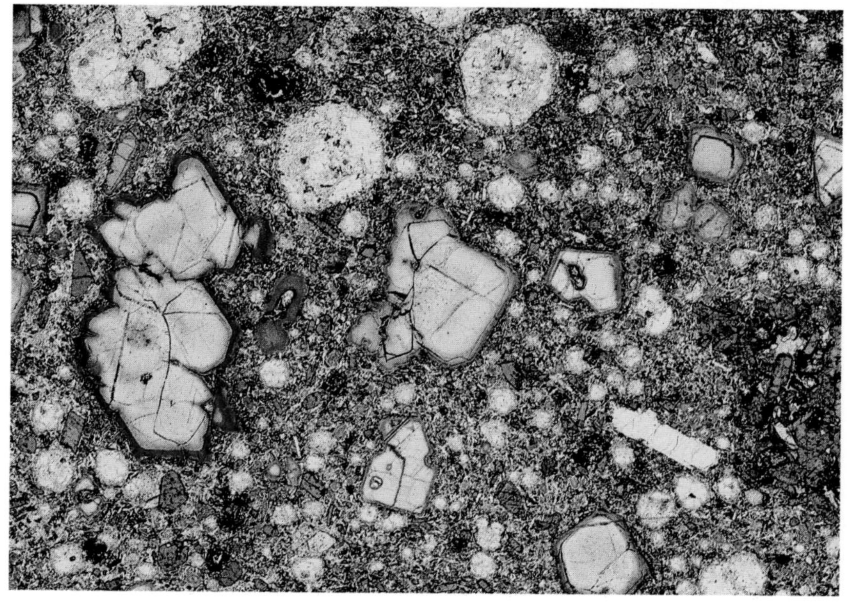

149

Pseudoleucite phonolite

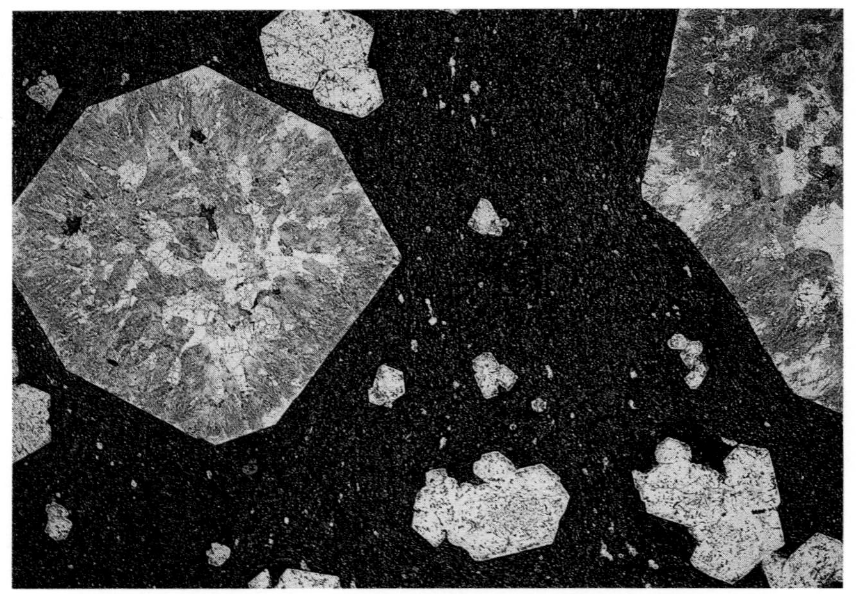

This term is used to describe a rock which contains phenocrysts having the shape of leucite crystals but composed of a pseudomorph aggregate of nepheline and K-feldspar. The groundmass consists of nepheline, alkali feldspar and alkali pyroxene.

The specimen illustrated shows two sizes of phenocrysts both having the outline of leucite or analcite crystals, i.e. eight- and six-sided sections. The two larger phenocrysts are predominantly brown in colour and this is mainly altered sanidine; the clear areas within them are mostly of nepheline and analcite. The clear area in the large pseudoleucite at the right edge of the field is analcite. The smaller phenocrysts differ in that they show very little brown alteration and have a higher concentration of needles of pyroxene. It appears that these may represent two generations of original leucites. The groundmass is an aggregate of minute crystals of nepheline, feldspar and pyroxene.

Pseudoleucite phonolite from Bearpaw Mountains, Montana, USA; magnification × 11, PPL and XPL.

150

Blairmorite

A very rare volcanic rock, known only from three or four localities, it is characterized by the presence of phenocrysts of analcite and sanidine in a groundmass of analcite, alkali feldspar and pyroxene. Melanite garnet is frequently present. Despite its rarity, it is included here because it is the only extrusive rock which has both analcite and garnet as phenocrysts. Its chemical composition is that of a sodium-rich phonolite but the nepheline which is present in a phonolite is here represented by analcite, and it has been suggested that the analcite is a primary mineral and not a replacement of leucite.

The photographs show three phenocrysts of sanidine (grey interference colours). Two of the phenocrysts are partially surrounded by analcite which has a slightly yellowish colour in the PPL view; the dark brown microphenocrysts are of melanite garnet; and the green crystals are of aegirine augite. The groundmass consists of laths of sanidine, equant analcite and pyroxene.

Blairmorite from Blairmore, Crow's Nest Pass, Alberta, Canada; magnification × 12, PPL and XPL.

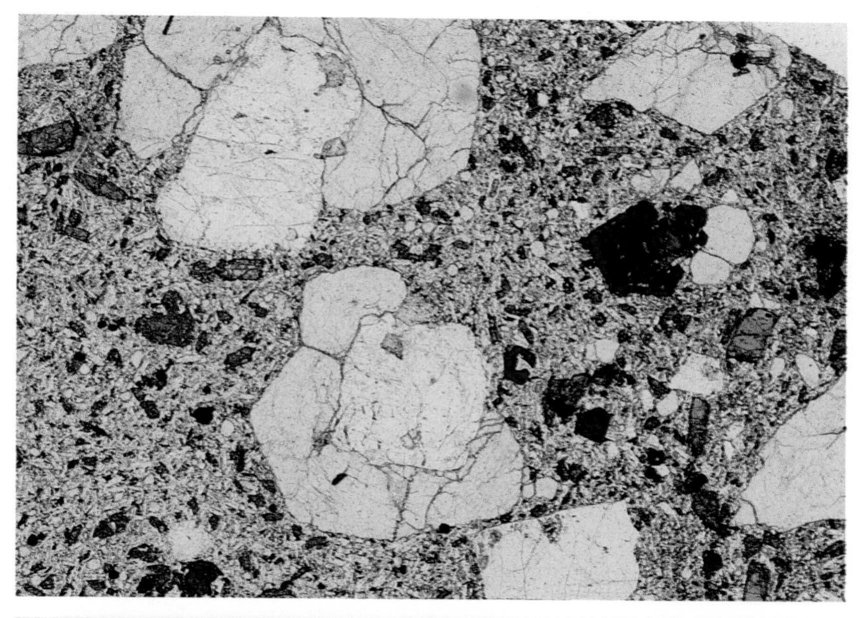

151

Nepheline syenite

This is the name used for a coarse-grained, felsic rock consisting essentially of alkali feldspar and nepheline with a small proportion of mafic minerals, usually alkali amphibole or pyroxene or both. It is the plutonic equivalent of *phonolite*. The medium-grained equivalent is best called *nepheline microsyenite*, though the special name *tinguaite* also exists for this.

The name *foyaite* has been used for nepheline syenites which have a trachytoid texture and the first two rocks illustrated fall in this category. The term *ditroite* is sometimes used for nepheline syenites in which the texture is subhedral granular, and the third rock illustrated could be so described.

We have chosen to illustrate this rock type by three XPL photographs, since the PPL views are dominated by colourless low relief minerals and hence are less informative. The first photograph shows interpenetrating tabular crystals of feldspar in which the interference colours are patchy. This is partly due to the fact that the crystals are microperthitic, and partly due to the presence of multiple twinning in both the sodium- and potassium-rich phases of the microperthite. A region of uniform grey interference colour can be seen just below the centre of the field. This is nepheline, as is a small triangular area towards the top right-hand part of the field. The small black triangular area just above the centre of the field and to the left is analcite. At the top left of the field are a few crystals of pyroxene showing a green interference colour, and an alkaline amphibole showing a very dark brown colour can be seen at the lower right of the centre of the field.

The second photograph shows tabular phenocrysts of microperthitic-feldspar and microphenocrysts of nepheline showing homogeneous grey interference colours – a large area just above and to the left of the centre of the field of view is nepheline. Between the phenocrysts is a trachytic-textured groundmass consisting of nepheline, feldspar, pyroxene and a few biotite crystals showing bright second-order interference colours. The magnification used here is such that we cannot easily distinguish the ferro-magnesian minerals from the one photograph. All the regions in which the interference colour is a uniform light or dark grey are of nepheline.

The third photograph is of a slightly coarser-grained rock than the other two, but here again the alkali feldspar and nepheline can be fairly easily distinguished by the homogeneous interference colour shown by the nepheline, in contrast to the microperthitic texture of the feldspars. A few crystals of nepheline are visible to the right of the centre of the field. The small coloured crystal to the right of the centre is biotite and in the bottom right corner a few crystals of sphene are visible.

First photograph: Nepheline syenite from Pilansberg, South Africa; magnification × 12, XPL.
Second photograph: Nepheline microsyenite from Barona, Portugal; magnification × 7, XPL.
Third photograph: Nepheline syenite from Langesundfjord, Norway; magnification × 11, XPL.

152

Malignite

This is the name given to a rock consisting essentially of pyroxene, alkali feldspar and nepheline in which the pyroxene is the dominant constituent (about 50 %) and the other two minerals are in approximately equal amounts. It can be considered to be a mesocratic variety of nepheline syenite.

The rock illustrated shows a large number of equant euhedral green pyroxene crystals. The clear crystals to the left of the centre, showing uniform interference colours are of nepheline, the other parts are of feldspar. The region at the right edge of the field is composed of a nepheline-feldspar intergrowth. A crystal of biotite is visible at the top left corner of the field.

Malignite from Shan-xi, China; magnification × 7, PPL and XPL.

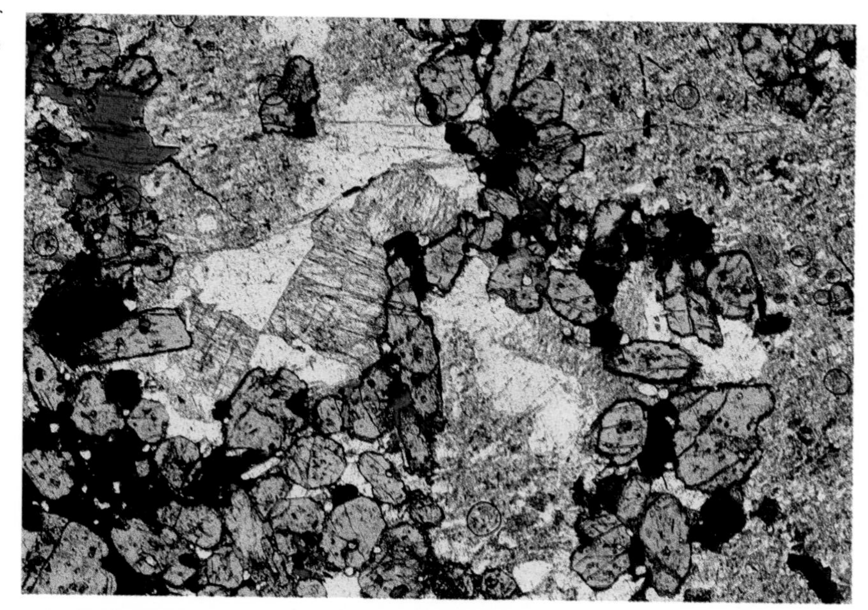

153
Sodalite syenite

A coarse-grained rock consisting essentially of sodalite and alkali feldspar with generally some nepheline also. Alkali amphiboles and pyroxene are invariably present.

The view illustrated here shows large euhedral sodalite phenocrysts, isotropic in the XPL view, amphibole and pyroxene in a finer-grained aggregate of alkali feldspar and nepheline. The alkali feldspar shows the same patchy extinction as seen in the nepheline syenites illustrated, whereas the nepheline has uniform interference colours – one nepheline crystal showing a pale grey interference colour can be seen adjacent to two sodalite crystals, in the left part of the view. Another clear nepheline crystal can be seen to the left of the large sodalite at the bottom edge of the field. The two lath-shaped isotropic crystals are also of sodalite, indicating a tabular habit of the crystals in this rock.

Sodalite syenite from Julianhaab, Greenland; magnification ×11, PPL and XPL.

154

Nephelinite

This is a fine-grained rock consisting essentially of nepheline and pyroxene without olivine (when olivine is present the name would be *olivine nephelinite*). If more than 50% of mafic minerals is present the term *melanephelinite* is used. *Ijolite* is the coarse equivalent of nephelinite.

The rock illustrated in the first two photographs consists mainly of pyroxene microphenocrysts in a finer-grained, seriate-textured aggregate of pyroxene, nepheline and iron ore. The pyroxenes show a pale greenish colour in the PPL photograph, whereas the nepheline crystals are smaller and are transparent. They show a rectangular or square shape, and one nepheline crystal is clearly visible near the top right corner of the field. The small patches of a mineral showing a fox-brown absorption colour are of biotite; iron ore is widely distributed throughout the rock. The third photograph shows a porphyritic nephelinite in which the phenocrysts are of nepheline and pyroxene in a groundmass which is extremely fine grained, and made of the same minerals. The only other mineral present in this rock is iron ore.

First and second photographs: Nephelinite from Mayotte, Comoro Islands, Indian Ocean; magnification × 53, PPL and XPL.
Third photograph: Nephelinite from Mayotte, Comoro Islands, Indian Ocean; magnification × 11, XPL.

155
Ijolite

An ijolite is a mesocratic coarse-grained rock consisting essentially of nepheline and clinopyroxene in approximately equal amounts. Frequently melanite garnet is present. This is the coarse-grained equivalent of *nephelinite*. Other names used for coarse-grained nepheline-pyroxene mixtures are *alkali pyroxenite*, *melteigite* and *urtite* (see **156**), these respectively being hypermelanic (< few per cent nepheline), melanocratic and leucocratic. The pyroxene can range in composition from sodic diopside to aegirine augite to aegirine to titanaugite. If titanaugite is the pyroxene in an alkali pyroxenite, then the name *jacupirangite* is used.

In the granular-textured specimen we have used to illustrate this rock type, the nepheline is recognized by its low relief in the PPL view and by the first-order grey interference colours in the XPL view. At the edges of the nepheline crystals, and in cracks within them, the mineral which has first-order pale yellow and white colours is cancrinite. The pyroxene in this rock is not strongly coloured but shows only a pale green absorption colour.

Ijolite from Alno, Sweden; magnification × 16, PPL and XPL.

A leucite-bearing variety of microijolite is shown in ***20****.*

156

Urtite

A coarse-grained, leucocratic rock consisting mainly of nepheline, but generally an alkali pyroxene and/or an amphibole is present in small amounts. It is more felsic than *ijolite*. (**155**)

The field of view is mainly occupied by nepheline, somewhat altered and showing a brown colour in the PPL view. The dark crystals in the field are of an alkali pyroxene considerably altered in parts. At the centre of the bottom edge of the field, an area which was probably pyroxene is almost entirely filled with fine-grained alteration products of the pyroxene, except for a group of clear crystals (PPL view) which are of apatite.

Urtite from Khibina, Kola peninsula, USSR; magnification ×11, PPL and XPL.

157
Basanite

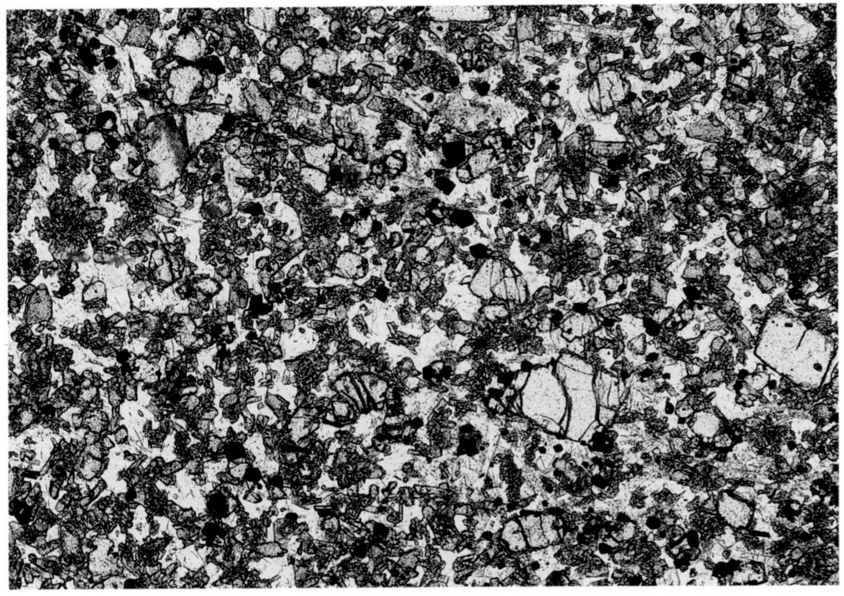

A fine-grained mesocratic rock containing essential olivine, augite, plagioclase feldspar and a feldspathoid, generally nepheline, with or without analcite or leucite[1]. The equivalent olivine-free rock is called *tephrite* (see **158**). The coarse-grained equivalent is known as *olivine theralite*.

This sample contains microphenocrysts of olivine which stand out in the PPL view because of their lack of colour and high relief. The groundmass is made up of pale brownish pyroxenes, scarce laths of plagioclase (e.g. at top right) and abundant poikilitic areas of nepheline (e.g. in the centre of the field of view and at the left side showing low grey interference colours). To the bottom left of the photograph the pyroxene is intergrown with a colourless mineral of lower refractive index than the nepheline and with almost zero birefringence; this has been identified as analcite by microprobe analysis. The presence of abundant nepheline and limited plagioclase in this rock is what distinguishes it from an olivine basalt.

Basanite from Jabal os Sawda, Libya; magnification × 27, PPL and XPL.

---

[1]The name *leucite basanite* should be used if leucite is present since *basanite* is generally taken to mean that nepheline is the feldspathoid.

158
Tephrite

A tephrite is a fine-grained mesocratic rock containing, in addition to plagioclase and pyroxene, nepheline or another feldspathoid[1]. Olivine is absent; if it is present, the rock is called a *basanite* (see **157**), even though *olivine-tephrite* would be more logical. *Theralite* is the coarse-grained equivalent of tephrite.

The sample we have used to illustrate this rock type is a hauyne tephrite. It contains phenocrysts of clinopyroxene and hauyne. The clinopyroxene crystals are euhedral and have a dark olive-green colour in the PPL view; both zoning and twinning can be seen in the XPL view. The hauyne phenocrysts are either blue or colourless and have rims of a darker blue or brown colour. The groundmass consists of lath shaped crystals of plagioclase, small round crystals of leucite, together with hauyne and clinopyroxene.

Hauyne tephrite from Monte Vulturi, near Malfi, Italy; magnification ×27, PPL and XPL.

[1] *Tephrite implies that nepheline is the feldspathoid; if another feldspathoid is present in place of nepheline, it prefixes the name, e.g.* leucite tephrite.

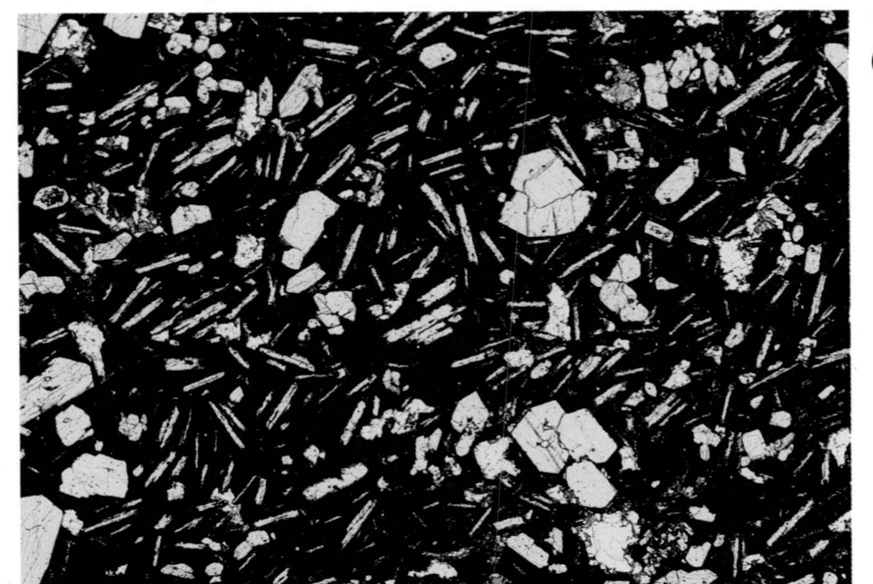

159
Olivine melilitite

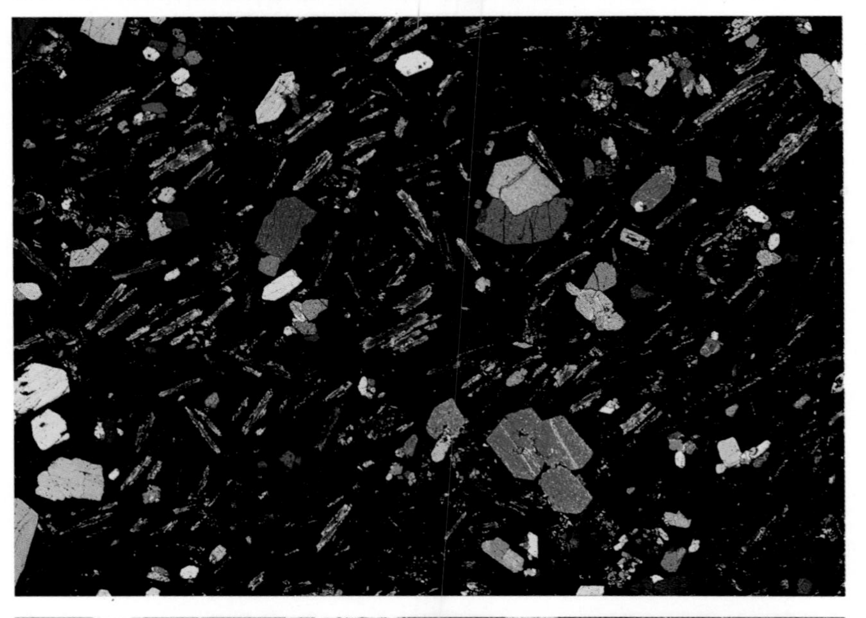

This rock is composed of essential olivine, melilite and pyroxene. Perovskite is a frequent accessory constituent and nepheline may be present. The name *melilite basalt* has been used for this type of rock but it is not appropriate since plagioclase is absent, its place being taken by melilite.

The sample we have illustrated contains very little pyroxene and so consists mainly of olivine and melilite crystals in a glassy groundmass. In the PPL view the rock appears to consist of olivine and plagioclase microphenocrysts, but it can be clearly seen from the XPL photograph that what one might take to be laths of plagioclase are in fact melilites. These are clearly identified by their anomalous interference colours. Closer inspection of the PPL view shows that many of the melilite laths have a line along the centre of crystals which is characteristic of melilite crystals. In addition to olivine and melilite, perovskite is fairly abundant in this rock. The third photograph is a high-magnification view of part of the field of view shown in the first two photographs, and shows a number of multiple-twinned perovskite crystals: in this photograph they are dark green in colour. This section must be slightly thin because the olivine crystals do not show colours as high on Newton's scale as expected from a mineral whose birefringence is between 0.035 and 0.052.

Olivine melilitite from Katunga, Uganda; magnification ×15 (first and second photos) PPL and XPL, ×72 (third photo), XPL.

160
Leucitite

This is the name given to an extrusive mesocratic rock consisting essentially of leucite and a clinopyroxene without olivine. If olivine is present the name used is *olivine leucitite*, and if the olivine and pyroxene dominate it is known as *ugandite*. The intrusive equivalent of a leucitite is sometimes called a *fergusite* (see **161**).

The sample we have illustrated shows, in the PPL view, clear phenocrysts and glomerocrysts of leucite and zoned olive-green crystals of pyroxene, set in a very fine-grained groundmass composed mainly of these two minerals and glass. There is no evidence of any other minerals such as nepheline or feldspar.

Leucitite from Celebes; magnification × 12, PPL and XPL.

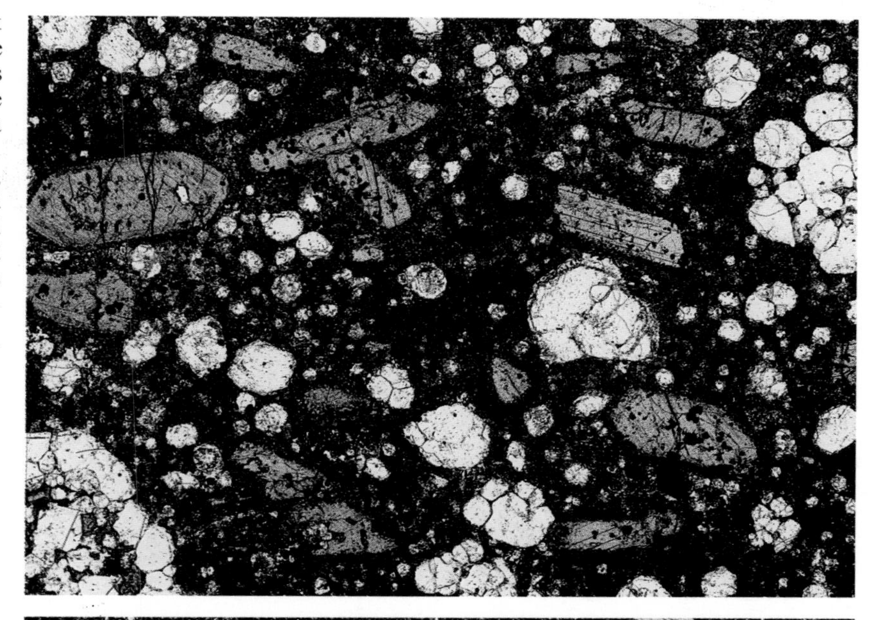

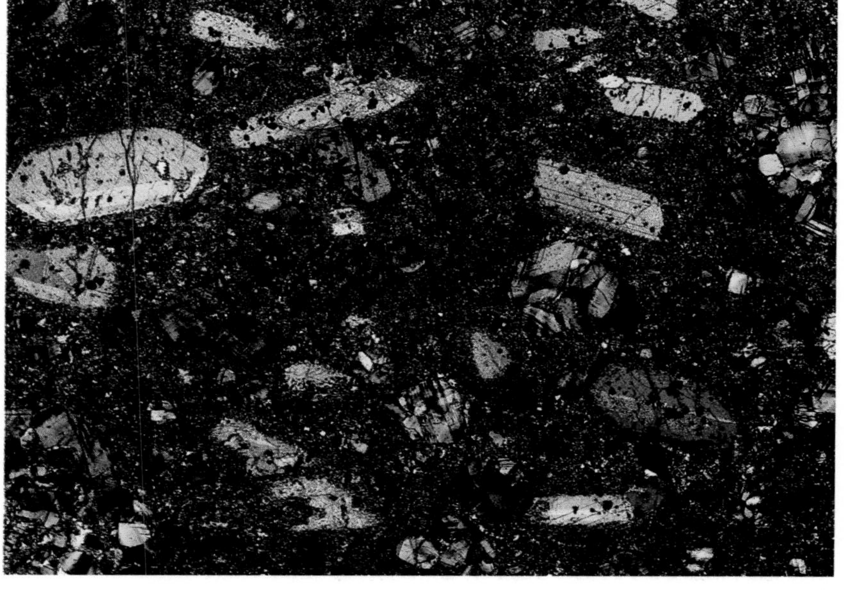

161

Fergusite

This name is used for the intrusive equivalent of leucitite (**160**) and it consists of pyroxene and leucite, or pseudoleucite. [In many rocks, crystals having the form of leucite are composed mainly of nepheline and K-feldspar and these are known as pseudoleucites because they are presumed to be pseudomorphs after leucite (see **149**).] Chemically the rock resembles *shonkinite* (**141**) but, being more silica-undersaturated, leucite takes the place of K-feldspar.

This sample shows subhedral to euhedral crystals of a greenish pyroxene and a nearly colourless mineral. This colourless material is made up of rounded aggregates of crystals, some of which are undoubtedly leucites – others are possibly nepheline-feldspar intergrowths. A small amount of an opaque mineral is also visible, as are a few crystals of biotite.

Fergusite from north of Maria, South-west Celebes; magnification × 16, PPL and XPL.

162
Minette

A *minette* belongs to the lamprophyre group and consists of essential biotite and K-feldspar; augite and plagioclase may be present as accessories. *Vogesite* has the same minerals, except that amphibole takes the place of biotite. *Kersantite* and *spessartite* are the equivalents of minette and vogesite, except that plagioclase, rather than K-feldspar, is the essential feldspar.

The term *lamprophyre* is given to a diverse group of rocks which traditionally have been grouped together because of apparent similarities in texture, mineralogy and occurrence. They are porphyritic rocks, occurring typically in dykes and sills, and characterized by euhedral phenocrysts of ferro-magnesian minerals in a fine-grained groundmass containing the same minerals with feldspar or feldspathoids. Unlike most rocks with the same essential minerals, lamprophyres have an abundance of ferro-magnesian minerals and hence are melanocratic-mesocratic in colour index; calcite and chlorite are common in the groundmass.

The name *lamproite* (**165**, **166**, **167**) has been used as a group name for extrusive rocks rich in potassium and magnesium. These rocks may be considered as phlogopite-rich extrusive equivalents of potassic lamprophyres.

The specimen used to illustrate this rock was chosen for its lack of alteration, and a region in which there is a segregation of quartz and alkali feldspar has been photographed. The edges of the sanidine crystals are clearly defined against the clear quartz by the multitude of haematite crystals in the sanidine. The top half of the field is more typical of the general appearance of the thin section in that it consists mainly of sanidine, biotite and a carbonate mineral. The carbonate can be identified from the photograph as the small clear patches in PPL view, which show high interference colours in the XPL view. In addition to these phases, the rock also contains very small crystals of an alkaline amphibole, apatite and rutile.

Minette from Pendennis Point, Cornwall, England; magnification × 20, PPL and XPL.

163
Alnöite

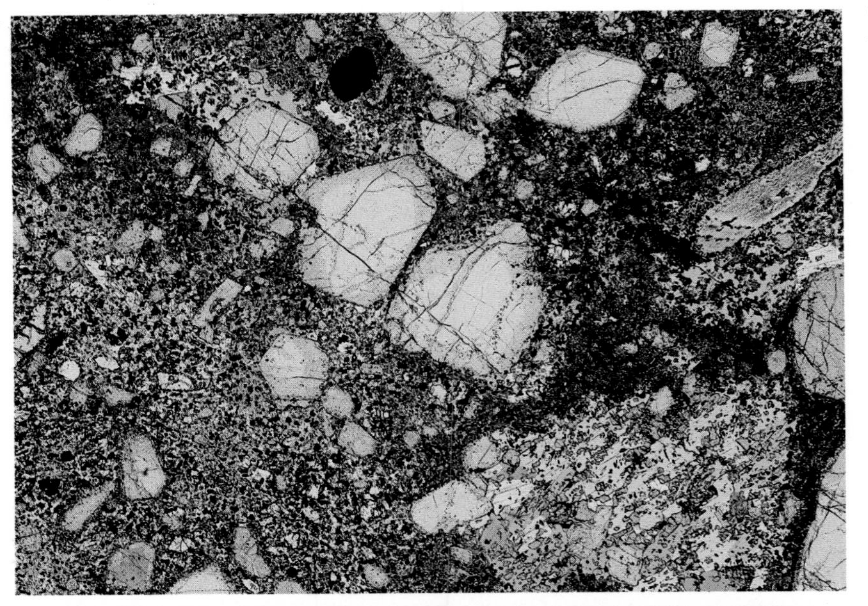

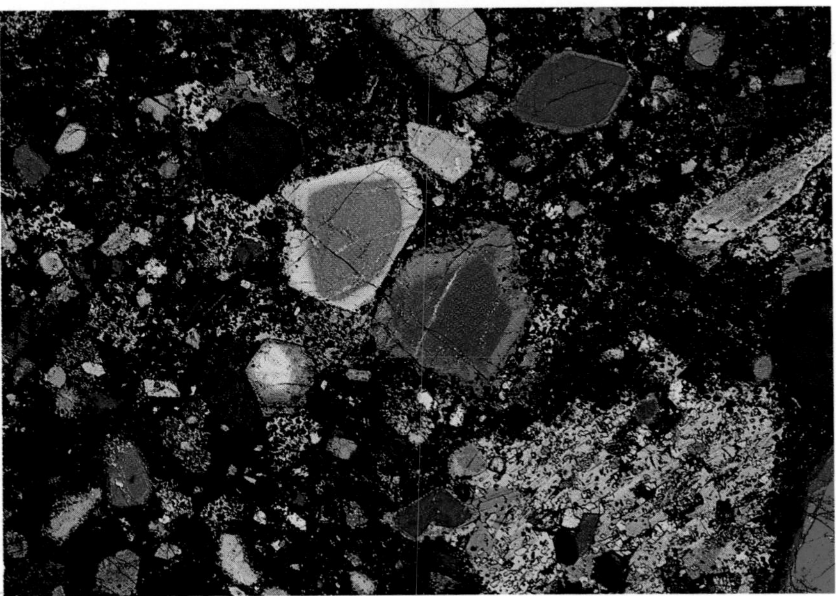

This rock belongs to the alkali lamprophyre group. It contains olivine, pyroxene and biotite, usually as phenocrysts, in a groundmass containing melilite; a carbonate mineral is commonly present also. Monticellite may be present. *Camptonite* consists of essential plagioclase and brown amphibole; titanaugite may be present, possibly in excess of the amphibole; a little analcite may occupy the interstices. *Monchiquite* has essential olivine, titanaugite and brown hornblende in a matrix of analcite and possibly some biotite. *Fourchite* (of which an unusual textured variety is shown in **355** and **70**) is the same but without olivine.

In the specimen illustrated here most of the phenocrysts in the centre of the field of view are zoned clinopyroxenes. A few microphenocrysts of olivine are present. At the bottom right of the field of view is a large area which is an aggregate of crystals of biotite intergrown with pyroxene. In the groundmass are poikilitic regions showing a pale brown colour and these are also of biotite. A pale brown elongated crystal of mica can be seen towards the right edge of the field.

Much of the clear groundmass material seen in the PPL view is melilite but because of its very low anomalous interference colour it is not easily identified in the XPL view at this magnification. The dark brown crystals are mainly perovskite.

Alnöite from Oka, Quebec; magnification × 15, PPL and XPL.

164
Mafurite

A rare volcanic rock type which consists mainly of clinopyroxene, kalsilite and olivine. Other minerals which may be present are leucite, melilite, perovskite and phlogopite. If the rock had more silica, the kalsilite would be present as leucite and the rock would be a leucitite or olivine leucitite. If glass is present the term *katungite* is used.

The view illustrated was chosen because there is a segregation of kalsilite crystals in which their square outline can be seen; it is not possible to distinguish these optically from nepheline. The phenocrysts consist of olivine and pyroxene, the latter has a brownish-green colour and good cleavage visible in some crystals. The yellow and light brown patches are of phlogopite. The groundmass of the rock is made up mainly of needles of clinopyroxene subophitically enclosed in kalsilite – no obvious leucite or melilite crystals can be detected in this view. The opaque mineral is an iron oxide but some crystals having a deep brown colour and showing multiple twinning are perovskite.

Mafurite from Bunyarugaru, Uganda; magnification × 24, PPL and XPL.

165
Fitzroyite

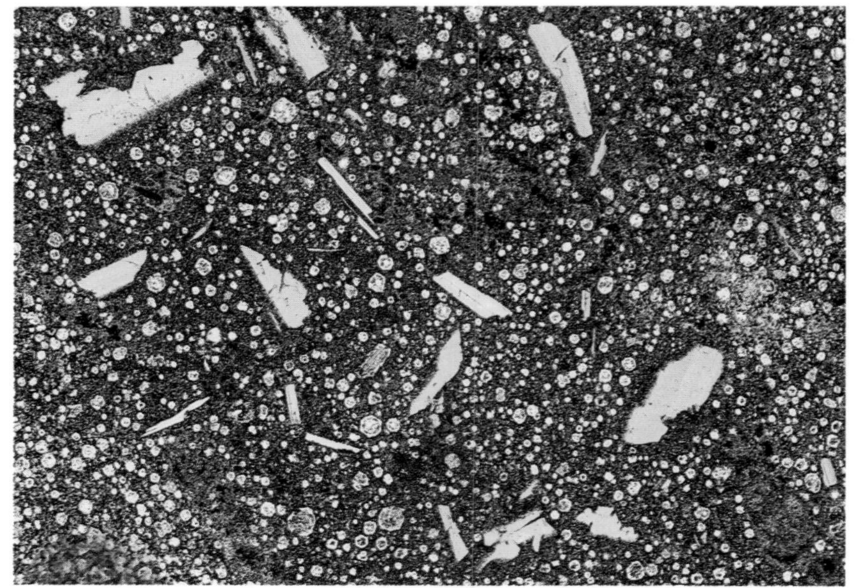

A potassium-rich lava which consists essentially of phlogopite and leucite. This is a member of the group of rocks known as lamproites (cf. *wyomingite* **166** and *madupite* **167**, see p. 133).

The photographs show microphenocrysts of phlogopite set in a groundmass which consists of equant, euhedral leucites and glass. The phlogopite in this rock is unusual in that multiple twinning is well developed. This can even be detected in the PPL view because of the slight difference in absorption colour of the different orientations in the twinned crystals. The larger leucite crystals show the characteristic zonal arrangement of inclusions of glass. In the groundmass are tiny brown elongated crystals originally thought to be rutile, but shown later to be priderite $(K,Ba)_3(Ti,Fe)_{16}O_{32}$ (see **167**).

Fitzroyite from Howes Hill, West Kimberley, Australia; magnification × 19, PPL and XPL.

166
Wyomingite

A potassium-rich lava consisting of phlogopite, diopside and leucite, in that order of relative amounts. It is one member of a group of rocks known as lamproites (cf. fitzroyite **165**). The term *phlogopite leucitite* would be preferable (see also **167**).

The sample illustrated shows microphenocrysts of phlogopite and diopside in a groundmass consisting of columnar crystals of the same two minerals in trachytic texture and also equant leucites. The phlogopite crystals are pale brown in the PPL view, but because of a strong preferred orientation in this section we do not see the full range of absorption colours. Diopside microphenocrysts have a greyish colour and one crystal near to the left edge of the field of view can be clearly seen because it is larger than most of the other diopside crystals.

In the groundmass, the small round white crystals are of leucite but these appear almost black in the XPL view. Under high magnification, multiple twinning can be detected in a few of the larger leucite crystals and this helps to confirm their identity.

Wyomingite from Leucite Hills, Wyoming; magnification ×28, PPL and XPL.

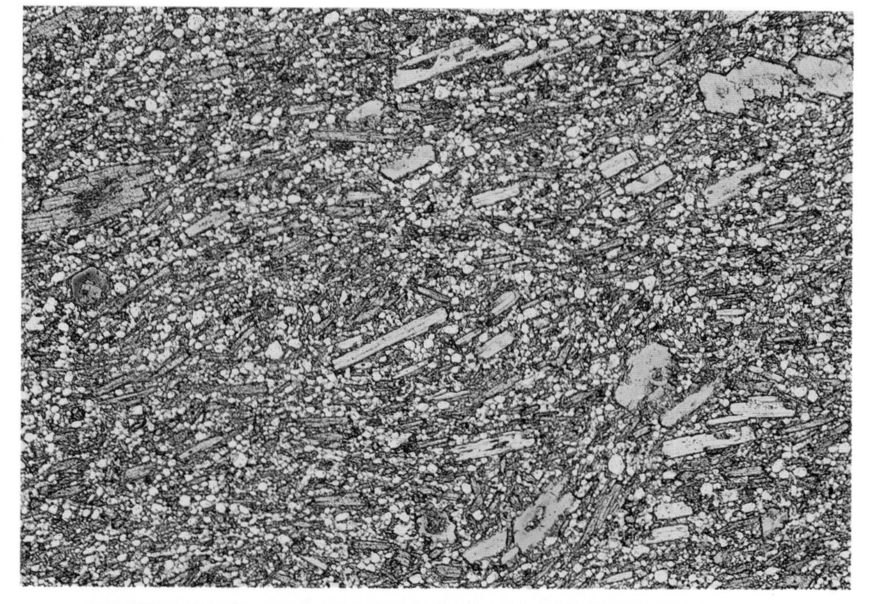

167
Madupite

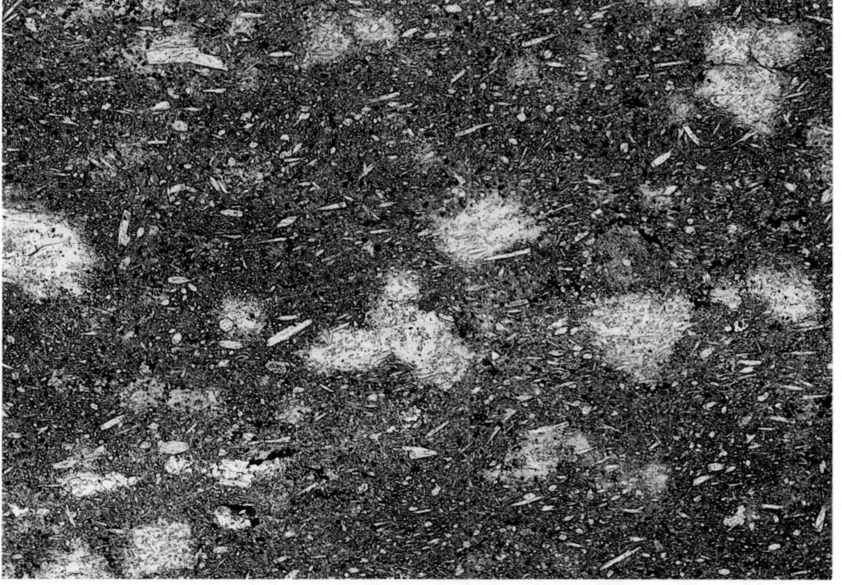

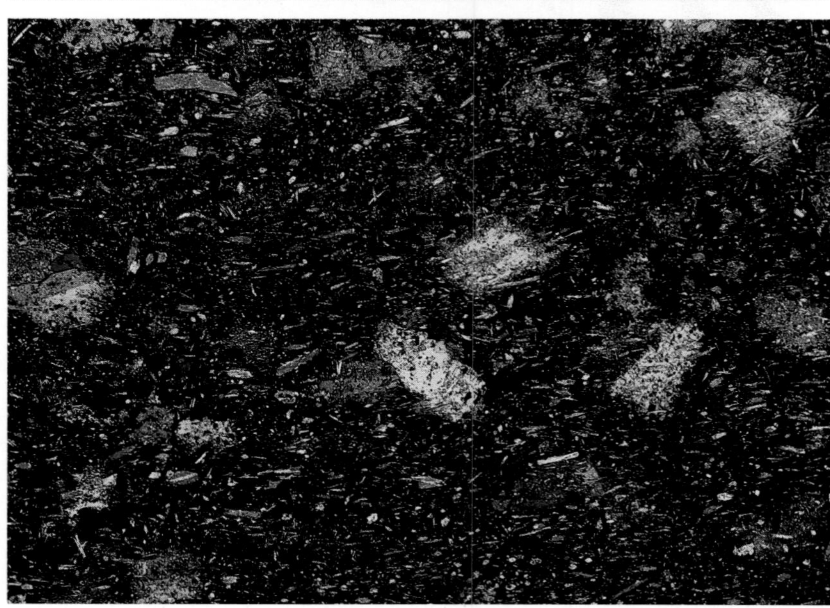

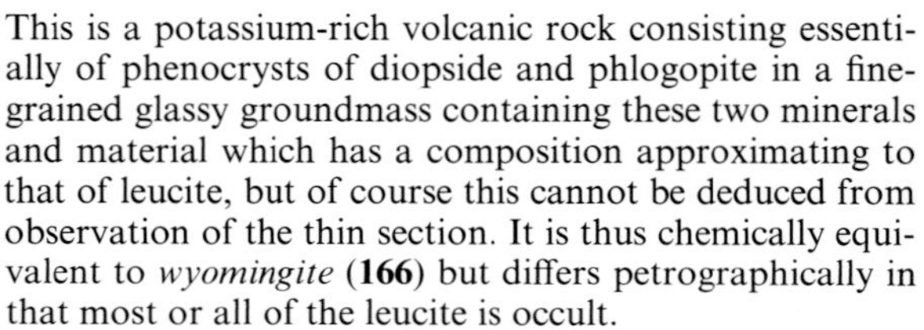

This is a potassium-rich volcanic rock consisting essentially of phenocrysts of diopside and phlogopite in a fine-grained glassy groundmass containing these two minerals and material which has a composition approximating to that of leucite, but of course this cannot be deduced from observation of the thin section. It is thus chemically equivalent to *wyomingite* (**166**) but differs petrographically in that most or all of the leucite is occult.

The rock illustrated shows phlogopite crystals ophitically enclosing small columnar crystals of diopside. Phlogopite also occurs in the groundmass with columnar trachytic diopside and careful comparison of the two photographs reveals clear equant crystals in the PPL view which appear isotropic in the XPL view. These are leucite crystals, an aggregate of which can be seen towards the lower left of the field of view. Small grains which appear black in the PPL view are in fact dark brown, and are of the rare mineral priderite (see p. 136).

Madupite from Leucite Hills, Wyoming; magnification ×16, PPL and XPL.

168
Carbonatite

This name covers a considerable variety of different volcanic and intrusive igneous rock types whose main constituent (greater than 50%) is a carbonate mineral. The most common carbonatite is called *sovite* which is a calcite carbonatite. Others contain dolomite, or siderite or alkali carbonate. The texture is generally granular but rarely may be trachytic or comb layered.

The specimen we have chosen to illustrate is a carbonatite composed of more than 90% calcite with minor amounts of apatite, pyroxene, monticellite, mica and an opaque mineral which we have not identified.

Carbonatite from Oka complex, Quebec, Canada; magnification × 7, PPL and XPL.

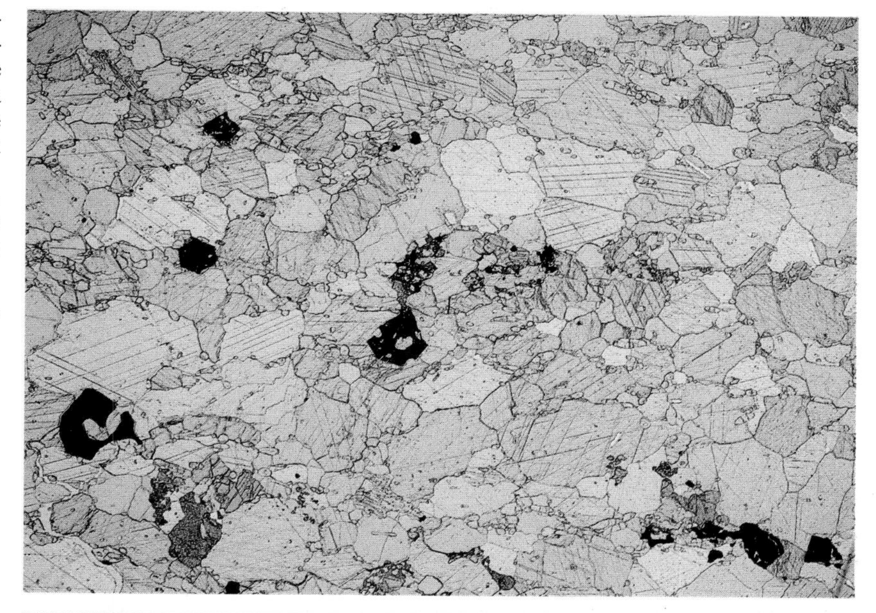

169
Chondrite
(meteorite)

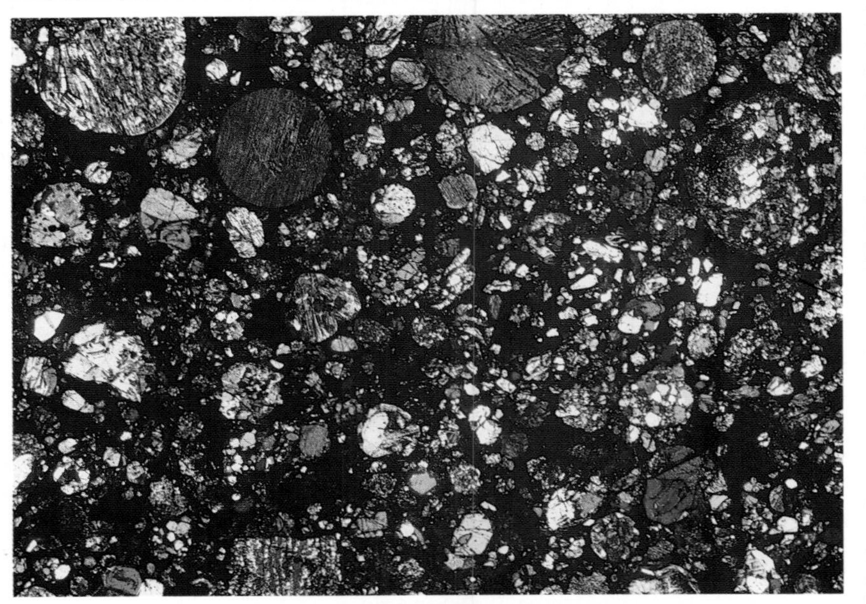

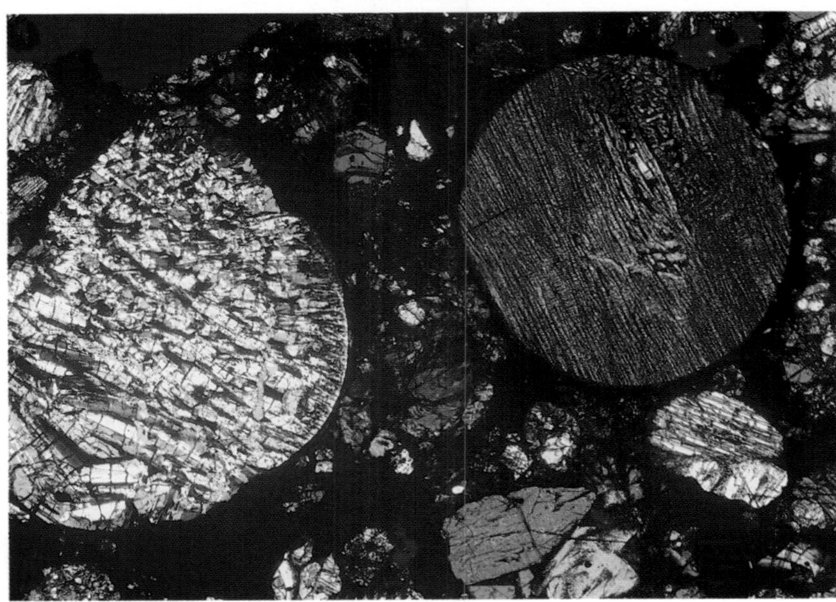

Meteorites are classed as stones, stony-irons or irons depending on the proportion of silicates to nickel-iron alloy. The stones are mainly composed of olivine and orthopyroxene and are subdivided into chondrites and achondrites according to the presence or absence of chondrules. Chondrules are spherical objects, with an average diameter of 1 mm, and are composed of olivine or pyroxene or both. We have illustrated two chondrites and one achondrite (**170**). Most authorities on meteorites prefer to consider chondrites as metamorphic rocks, though the chondrules may be of igneous origin.

The first photograph is a PPL view of an olivine – hypersthene chondrite which is the commonest type of chondrite. In the thin section only the one chondrule (which we have illustrated) is present. It consists of radiating crystals of orthopyroxene. The rest of the field is made up of an aggregate of orthopyroxene and olivine and some opaque regions. The opaque regions are mostly of a metal phase with some sulphides but these cannot be distinguished in transmitted light.

The second photograph is an XPL view of a chrondrite in which the chondrules are unusually well developed and can be seen to have different mineralogy and texture. One small chondrule, just above the centre of the field of view, showing a yellow birefringence colour with black lamellae is composed of olivine with lamellae of glass.

The third photograph is a higher magnification XPL view of the top left corner of the second photograph. The chondrule at the left of the field consists of bladed twinned crystals of clinobronzite. Because of the twinning and low birefringence it could be mistaken in a photograph, for plagioclase. This chondrule appears to have been broken at some stage since in section it is not a complete circle. The other large chondrule is also composed of pyroxene but is too fine grained for optical identification. Three separate olivine crystals are visible at the bottom of the field of view.

First photograph: Chondrite from Bruderheim, Alberta, Canada; magnification × 28, PPL.
Second and third photographs: Chondrite from Prairie Dog Creek, Kansas, USA; magnification × 16 (second photo), XPL, × 43 (third photo), XPL.

170

Achondrite

meteorite

Achondrites are stony meteorites (see p. 140) in which chondrules are lacking and a nickel iron phase is either absent or present in very small amounts. These are relatively rare meteorites.

The photograph shows coarse-grained clasts or fragments in a brecciated matrix. This specimen is classed as a *eucrite* and it consists mainly of bytownite and a pigeonitic pyroxene together with an augite. To the left of the centre of the field is a fragment consisting of an intergrowth of plagioclase and clinopyroxene. The plagioclase contains trails of minute inclusions unevenly distributed within the crystals and it is these inclusions which cause the brown colours in the crystals in the PPL view. The clinopyroxene is relatively iron-rich and a fine lamellar structure can just be detected in the XPL photograph. The rest of the field is occupied by the same minerals but of much finer grain size in some parts. The pyroxene crystals are almost black due to a very high concentration of an opaque mineral, probably magnetite. There is no olivine in this specimen.

The name *eucrite* is also used for a terrestrial gabbroic rock consisting of a calcic plagioclase (An_{70-90}) and a clinopyroxene. Some of the achondrites have textures similar to those of terrestrial gabbros.

Pyroxene-plagioclase achondrite from Stannern, Czechoslovakia, (observed fall, 1808); magnification × 14, PPL and XPL.

Appendix

Preparation of a thin section of rock

It is sometimes believed that complex and expensive equipment is required for making thin sections of rock of standard thickness of 0.03 mm but as the following instructions indicate this is not the case. Thin sections can be made by the amateur with a little patience and perseverance. If a diamond saw is available to cut a slab of rock 1–2 mm in thickness the process is considerably speeded up, but alternatively a chip of rock not more than 8–10 mm in thickness from which to make a thin section can usually be broken from a hand specimen with a small hammer.

The operations required to prepare a thin section after obtaining the fragment of rock are set out below.

Using 100 micron particle size (120 grade) carborundum abrasive, one surface of the rock fragment is ground flat on a piece of glass measuring about 30 cm × 30 cm and up to 1 cm in thickness; ordinary window glass is satisfactory if thicker glass is not available. Only a small amount of carborundum (half a teaspoonful), just moistened with water, is used for grinding. If too much water is present the carborundum tends to extrude from underneath the rock, and in consequence is much less effective for grinding.

After grinding with a rotary movement for about half a minute the noise of the grinding changes because the carborundum grains lose their sharp cutting edges. The glass plate is washed clean and a fresh slurry of carborundum made on the plate. The time spent on grinding a flat surface will of course depend on how irregular the surface of the rock chip was to begin with.

When the surface of the rock is flat the sample should be thoroughly cleaned with a jet of water before grinding with a finer grade of carborundum. The second stage of grinding should be carried out with 60 micron size (220 grade) carborundum and two periods of grinding for about a minute each with a fresh quantity of carborundum is all that is required at this stage.

After washing, a final grinding of one surface is made for about a minute with 12 micron size carborundum (3F grade). Again, after cleaning, the rock sample may be polished using cerium oxide (0.8 micron size) but this is not essential.

The next stage is to glue the smooth surface of the rock on to a microscope slide in one of two ways. It can be achieved by using a cold-setting epoxy resin which usually consists of two fluids which must be thoroughly mixed. The maker's instructions for using these should be followed carefully because these materials should not be allowed to come in contact with the skin and the vapour should not be inhaled. The refractive index of epoxy resins vary but most are somewhat higher than the value of 1.54. For any work involving comparison of the refractive index of minerals with the mounting material the refractive index of the cold resin should be ascertained. The chief disadvantage of using an epoxy resin is that it is very difficult to remove, if, for example, it became necessary to transfer the rock chip to another glass slide.

The alternative method is to use a material known as Lakeside 70C cement,[1] which is supplied in short rods and must be melted on a hotplate. This material begins to soften about 85°C so a hotplate which reaches 100°C is quite suitable. A

---

[1] *Lakeside cement is the proprietary name for a material manufactured in the USA and marketed in the United Kingdom by Production Techniques Ltd, 11 Tavistock Road, Fleet, Hampshire.*

flat piece of aluminium or steel placed on a gas stove or on the element of an electric cooker at very low heat can be used for this stage, if no electric hotplate is available. A glass microscope slide and the rock specimen should both be heated on the hotplate until they are just too hot to touch, and some Lakeside cement melted on both the flat surface of the rock and the slide by touching the hot surfaces with the rod of Lakeside cement.

Whether the cold-setting epoxy resin or the Lakeside cement is used, the procedure is the same at this stage in that the flat surface of the rock chip must be attached to the glass slide with no air bubbles between the two surfaces. The rock chip is placed on the glass slide and with a slight pressure and circular movement the excess mounting material and air bubbles are squeezed out. The slide is then turned over to observe whether, between the rock and the slide, any air bubbles have been trapped; any bubbles must be gently extruded by pressure and, in the case of the Lakeside cement, this has to be done before the cement cools and becomes too viscous for the bubbles to escape easily. It can be reheated to render it fluid again. With the epoxy resin, since the hardening takes place over a period which depends on the variety, more time is available for extruding the air bubbles but in this case the sample should not be heated because this only speeds up the hardening process.

If a diamond saw is available the rock fragment can now be cut from its original thickness of 5–10 mm to about 1 mm, otherwise it must be ground by hand. Its thickness should be reduced to about 0.2 mm (200 microns) using 100 micron size carborundum; at this thickness it is possible to see through the transparent minerals. Carborundum of 60 micron size should be used to reduce the thickness from 0.2 mm to 0.1 mm and at this stage quartz and feldspars should show bright second-order interference colours when examined under crossed polars.

The final stage of grinding from 0.1 mm to 0.03 mm is accomplished using 12 micron size carborundum. This is the stage in the whole process of section making which requires the most skill. The grinding has to be done very carefully to ensure that the section is of uniform thickness over its whole area, otherwise the edges tend to be ground preferentially and become too thin. The slide must be examined between each stage of grinding to check on the uniform reduction of the interference colours.

In the making of thin sections, it is generally assumed that the rock will contain some quartz or feldspar. These show first-order grey and white interference colours in a thin section of standard thickness and neither should show a first-order yellow or red colour. Thus a thin section in which quartz or feldspar shows colours in Newton's scale higher than first-order white is too thick. In the rare instance in which no quartz or feldspar is present, e.g. in a peridotite or a carbonatite, the thickness of the section is very difficult to estimate and has to be judged by the appearance of other minerals; only an experienced thin-section maker can estimate the thickness in such cases.

It is usual to cover the section, either by painting the surface with a transparent cellulose lacquer, or better with a glass cover slip since lacquer tends to scratch easily. This is traditionally done using Canada balsam diluted in xylene but the process of heating the mixture at the correct temperature for the correct time requires some experience. We have found that it is quite satisfactory to fix the cover glass either by using the same epoxy resin which was used to attach the rock to the microscope slide or by using a clear lacquer painted or sprayed on to the surface of the rock. As in the process of fixing the rock to the microscope slide, care must be taken to ensure that no air or gas bubbles are trapped between the cover glass and the rock. This is particularly important if the material has been applied by a spray because some of the propellant may be dissolved in the clear lacquer. Any bubbles which are visible in the liquid after spraying should be allowed to burst before applying the cover slip. Only enough lacquer or Canada balsam to cover the slide with a thin layer of liquid should be applied.

The cover slip should touch the liquid on the slide at one end and be allowed to fall slowly on to the liquid. If any air bubbles are visible they can be extruded by gentle pressure on the cover glass. The excess lacquer or epoxy resin must be extruded to render it as thin as possible, otherwise the minerals cannot be brought into focus with a high-power lens because of the short working distance of lenses of magnification more than $\times 40$.

Finally when the mounting material has set hard, the excess can be scraped from round the edges of the cover glass using a razor blade or sharp knife.

References

Cox, K. G., Bell, J. D. and Pankhurst, R. J., 1979, *The Interpretation of Igneous Rocks*. Allen and Unwin, London.
Hatch, F. H., Wells, A. K. and Wells, M. K., 1972, *Petrology of the Igneous Rocks*. Allen and Unwin, London.
Holmes, A., 1920, *The Nomenclature of Petrology*. Hafner, New York.
Holmes, A., 1921, *Petrographic Methods and Calculations*. Murby and Co., London.
Iddings, J. P., 1909, *Igneous Rocks*. Wiley, New York.
Johannsen, A., 1931, *A Descriptive Petrography of the Igneous Rocks*. University of Chicago Press.
Niggli, P., 1954, *Rocks and Mineral Deposits*. Freeman, San Francisco.
Nockolds, S. R., Knox, R. W. O'B., and Chinner, G. A., 1978, *Petrology for Students*. Cambridge University Press.
Wilkinson, J. F. G., 1968, The petrography of basaltic rocks: in *Basalts*, vol. 1, 163–214. Interscience, New York.

Index

All references refer to page numbers. Page numbers in *italics* refer to additional photographs of rocks in part 1 where detailed descriptions of them are included in part 2.

Index